LUNAR LANDING and RETURN

A Simplified Physics & Mathematics Investigation

The Apollo 11 Saga

by

Donald C. Lundy

ISBN: 0-75961-858-5

This book is printed on acid free paper.

1stBooks - rev. 1/4/02

Introduction

It is the intended purpose of this book to take the reader on a hypothetical trip to the moon and back, based upon the original Apollo 11 journey. You will follow the same course and schedule, based upon the mathematics of the original flight, and return to earth, landing in the Pacific Ocean in the proximity of the original crew landing site. The mathematics will be non-calculus, involving high school level algebra and trigonometry. You will be introduced to some topics in physics that are necessary for application of the applied mathematics to space travel. If you have a good background in physics and mathematics, you can skip much of the review offered as introductory to these topics. All review topics are followed with questions and problems related to the subject matter covered. Answers and problem solutions are provided for all given problems in the Self Test exercises.

The first chapter is a brief, historical, factual discussion of the astronomy of our solar system which includes the earth and its moon. If you are to travel through space to distant planetary objects, you must have some concept of their distances and motion. Most of this discussion has to do with planet earth and its moon and the interplanetary forces that bind them together in space and to the sun, the center of our immediate universe. As man will be limited in his ability to explore space beyond our solar system for years to come, it is beyond the purpose of this book to discuss galactic travel.

There is an extensive discussion of the Basic Language of computers in Chapter VII. The invention of the computer made space travel possible, because the mathematics of space travel and navigation require computer analogues of space trajectory. On-board computers control the engine thrust of landing space vehicles to bring them safely to ground landing. If you have not used Basic computer language before, you will find it a new means of doing mathematical work through the basic electronic network of your computer. If you do not have BASIC, you can still do most of the work presented in this book.

Before you begin to seriously read and do the book exercises, it will be to your advantage to leaf through the book from beginning to end. This will give you some idea as to where the subjects related to rockets and rocketry begin. Also, note that the Apollo sequence begins with Chapter VIII. If you have had a year of physics or are presently taking the course, you may find that you can skip most of the introductory material up to rocketry. For most readers, doing the introductory Self Test exercises will make it much easier to understand the information on rockets.

GOOD LUCK!

TABLE OF CONTENTS

PART I – Astronomy

PART II – Physics

PART III – The Apollo 11 Mission

PICTURES and ILLUSTRATIONS

PART I – Astronomy

Chapter I

The Moon and Beyond

The Greatest Adventure of All

On July 16, 1969, three great Americans left earth bound for the moon on the largest rocket ship ever made by man. As their Saturn rocket faded fast from the view of loved ones and scientists waiting below, these astronauts, and all other Americans who watched their departure on television, knew their chances of safely returning to earth were not very good. Their mission was to fly to the moon, land and walk on its surface, at the same time collecting specimens of rocks and soil, and then to return to earth. They did all of those things and landed in the Pacific Ocean 195 hours and 18 minutes after having left earth. This was the Apollo 11 Mission and the astronauts were Neil Armstrong, Col. Edwin Aldrin and Lt. Col. Michael Collins. You should know that Armstrong and Aldrin very nearly crashed on the moon, as they had less than 30 seconds of fuel left when the lunar module came to rest on its surface.

The successful flight of Apollo 11 was preceded by years of planning on the part of scientists and engineers. There were many rocket failures and many people died in their attempt to develop new types of fuel for American rockets. There was much computer work done in advance in order to pre-develop flight plans, and to determine the time of firing and the amount of controlled engine thrust. Many rockets were fired at the moon and around the moon. Tens of thousands of pictures were taken of the moon's surface as part of surveying missions in order to predetermine safe landing areas. Very little was left to chance, as any error could be fatal to the crew.

Now it is your turn to vicariously follow in the footsteps of these three great men. You will mount Saturn V and make the landing on the moon and safely return as they did. However, in order to do this, you must first learn to apply the mathematics and physics that made the Apollo series of missions possible, ie., that predetermines the flight, the acceleration and deceleration of the rocket. Without this knowledge, you can neither get off the ground or safely land your ship once in orbit. Furthermore, you need to have some idea of navigation and flight patterns as they apply to space and rendezvous with other moving objects-some as big as the moon. It will be important for you to learn how to write Basic programs that will provide flight data for your moving ship. All of these things had to be done in advance in order to assure the success of the Apollo 11 space exploration.

Space and the Universe

Our Universe is defined as "the totality of all existing things." If during the day you look into a cloudy sky, you could assume that only the earth and clouds make up your "Universe." If the clouds open up to permit the sun's rays through, then the sun is included in your universe. However, at night, all of the visible objects of the day have faded from view. When you look into the night sky, your eyes are overwhelmed by a whole new panorama slowly moving, brilliantly glowing, objects of varying density. If the moon appears over the horizon, you now have another object to add to this dimension. But, already you know that what you are seeing goes on and on and that trillions of similar objects are beyond the scope of view. But what is it that fills in the vast emptiness between all of these objects? If you answered "space" you are correct. Space is defined as the unlimited expanse in all directions in which material objects exits. Space is the substance that fills in the distance between all of these objects that you see during the day and at night.

If you wish to travel to any one of these bright objects that you see in the night sky, you must travel through space in order to get there. How wonderful it would be to be able to travel to these distant places in our universe. Have you ever used your imagination to project yourself to the moon or beyond? Using your mind, you can be there in a few seconds. If it is a star you are looking at, then you need to know it is probably much larger than our Sun, is a boiling inferno and has energy so intense that it would instantly vaporize your body. But suppose the star is surrounded by planets similar to earth. How wonderful it would be to be able to communicate with the inhabitants of such planets. Even if you could send a message to these alien beings, it would take many years for the signal to arrive there and the same number of years for their reply to return. In all probability, the sender of the original message would be dead before the return of the reply. In 1960, the Radio Astronomy Observatory at Green Bank, West Virginia, aimed its 26 meter radio dish at Tau Ceti deep in space hoping to receive messages from other intelligent life forms in that area of the galaxy. Later on, other countries made similar attempts, but all were unsuccessful. Man's first quest will be that of successfully communicating with extraterrestrial beings from other distant worlds in his own galaxy. His greatest achievement will come when he is able to invent the technology that will permit his travel to these very distant places.

Man is already well along in his conquest of space. He has sent many probes, some of them manned space probes, around the moon. Already his deep space probes have landed on the planets Mars and Venus. The Galileo spacecraft has sent images of ice on Europa, a moon of Jupitar, back to earth. The Pathfinder bounced to a soft stop on Mars and remitted over 9,500 images of the Martian surface back to earth. And most important of all, man has walked on the surface of the moon, not once, but three different times and lived to return to earth. So, he is already well along in his exploration of space. Now it is your turn to take this first great step into the future.

Planet Earth and Its Sun

Ancient man, awed by the intensity of the sun, worshipped it and made the sun his god. The sun gave him warmth and protection by day. At night he cowered in the coldness and darkness of its absence. Because of its apparent motion from horizon to horizon, he naturally assumed earth to be stationary, and the sun to move about earth in some unknown manner. It was thousands of years later that he could finally understand the enormous energy of the sun, and that he was a traveler riding through space on a planet that was located 93 million miles away from it.

How old are you? Assume that you are seventeen years old. Since birth you have been an inhabitant of this planet. Your planet has been circling the sun for 4.6 billion years and will one day be consumed by the sun when it becomes a Super Nova. You have been traveling through space for seventeen years. You can calculate this distance if you can calculate the circumference of a circle. The calculation is as follows, but don't let yourself be bothered by the units or powers of ten. There will be plenty of time to become familiar with both.

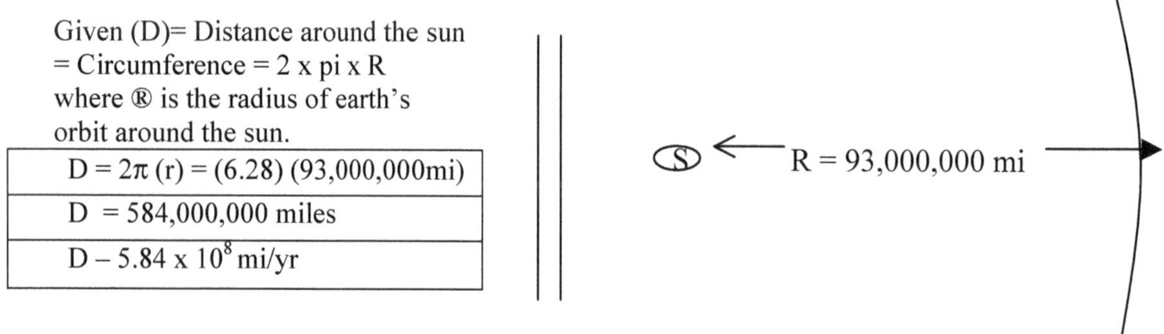

Given (D)= Distance around the sun = Circumference = 2 x pi x R where ® is the radius of earth's orbit around the sun.

$D = 2\pi (r) = (6.28)(93{,}000{,}000mi)$
D = 584,000,000 miles
$D - 5.84 \times 10^8$ mi/yr

Each year of your life, you have traveled 584 million miles through space. Use your calculator to calculate the total distance you have traveled in your lifetime. This answer should be 9.928 billion miles.

$$D(17\ yrs) = (17\ yrs)\ (5.84 \times 10^8\ mi/yr) = 9.928 \times 10^9\ miles$$

Even if earth were not moving around the sun, you would be a space traveler. In what other manner does earth move in space? Each day earth spins or rotates on its axis through one complete revolution. Even though you are not aware of this circular motion, you move with earth through a given distance each 24 hours. This motion is the same as that of an ant riding on a spinning top. If you spin the top fast enough, the ant will slide off the top. Why? Have you ever ridden on a giant turnstyle at an amusement park? Did you find it hard to hold on at higher speeds? If this motion is similar to that of earth, why do you not fall from earth in the same manner? If you answered "gravity," you gave the correct answer. Earth's gravity is 32 ft/sec-sec or 9.8 meters/sec-sec. These two numbers will become very important to you as you progress through the course.

Earth Facts

The best average measurement of the diameter of the earth is 7917.78 miles. Bear in mind that the earth is somewhat pair shaped and not quite perfectly round. How far do you rotate through space in one day if you are located at the equator? In making this calculation, use radius of the earth to be 4000 miles. The calculation is the same as that for your trip around the sun, except that the radius will be that of earth. You make this calculation. The correct answer is 25,120 mi/day. Divide this number by 24 hours and you will have calculated the rotational velocity of earth about its axis. This number is 1046.7 mi/hr. Do you travel a greater rotational distance per day or a greater circular distance per day as earth moves around the sun? Which motion is greater? Divide 584 million miles by 365.25 days per year. Was your answer 1,598,905 miles/day? How many times greater is this number than your rotational number? If you correctly divided the smaller distance into the larger distance, your answer is 63.65.

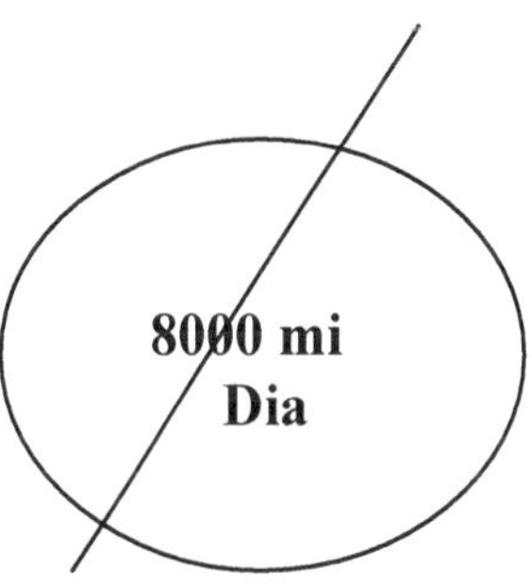

Planet Earth

There are a few other facts about our planet earth that will be of interest to you, and useful in making future calculations on or near the surface of earth. The earth's mass is 5.98×10^{24} kg which equals 6.58×10^{21} tons of weight. Such a large quantity of matter below the earth's mantle and core, create a gravitational pull that holds all matter to its surface. It is this great force that keeps all of your water and atmosphere from escaping into space. Were it not for gravity, this would have happened eons ago and mother earth would be a dead planet. About half of all of earth's atmosphere is contained in a very thin layer that envelops earth to an altitude of 3.5 miles. 99% of the earth's atmosphere exists in an area above earth that is only 21.88 miles deep. The troposphere is the lowest 10 mi. of atmosphere. Few aircraft can fly higher than this elevation. The stratosphere begins at this point and extends to about 50 mi. above earth. The exosphere includes the very outer limit of atmosphere which extends into the vacuum of space.

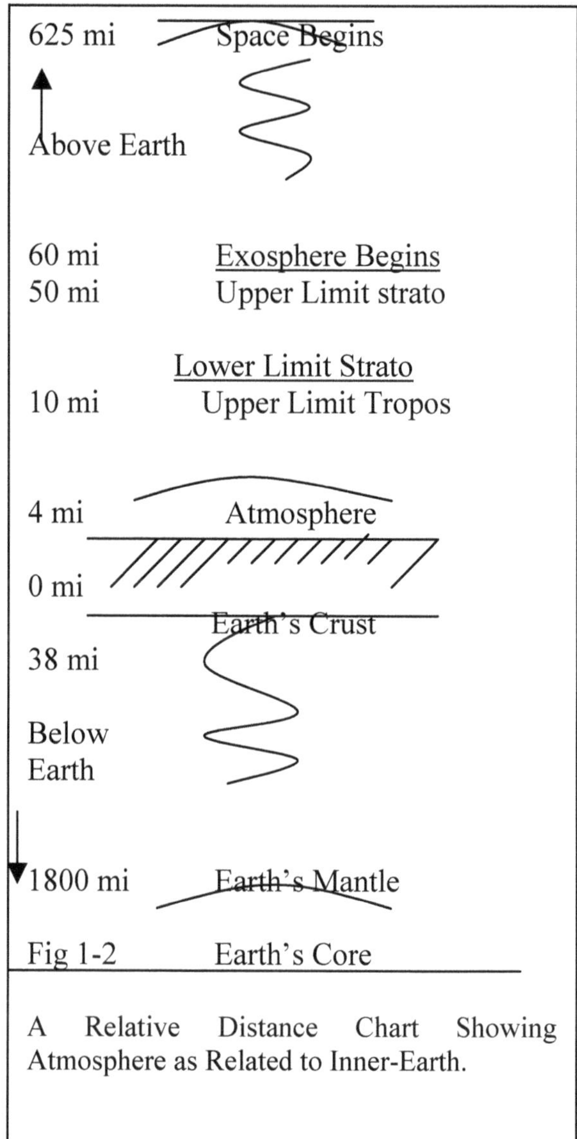

Fig 1-2

A Relative Distance Chart Showing Atmosphere as Related to Inner-Earth.

Exercise 1

Self Test

1. Name the three astronauts that made up the Apollo 11 crew.
2. If the Apollo 11 crew was launched at 9:32 AM Eastern Daylight Savings time, after how many days did they return to earth? At what time on this day did they return (Show how you arrived at your answer.)
3. Space travel poses many dangers to long term space travelers. List as many of these conditions as come to mind.
4. Astronomically define "universe" and "space".
5. Assuming it is possible to send messages into deep space at the speed of light, at what speed can these messages be returned to earth from outer space?
6. What total distance in miles have you traveled in 17 years due to the earth's rotation about its axis and its revolving about the sun? (Show your work in making this calculation.)
7. When you dive from a diving board, with what acceleration do you approach the water? (Use a number for your answer.)
8. What additional function of gravity has preserved life on earth? For this answer, name the substances.
9. What important quality about earth keeps you from falling from its surface into outer space? (Gravity is not the desired answer.)
10. What is man's first, greatest quest?
11. Looking at the chart on page (5), at what level above earth does space begin? (b) What is the approximate limit in elevation where life could be expected to survive on earth?
12. In which direction does the earth rotate? How can you be certain of your answer?

Earth's Moon

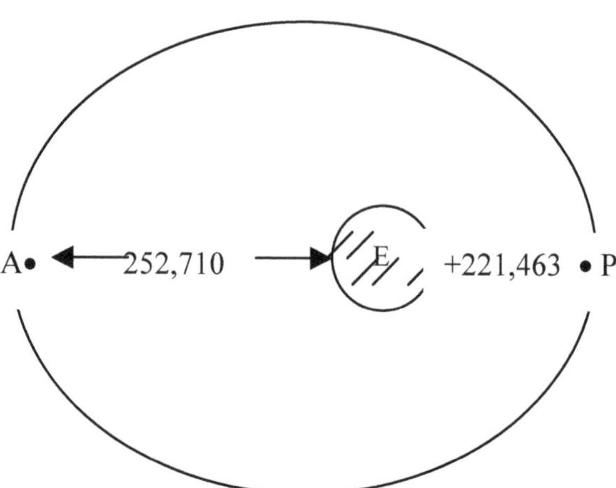

To ancient man, the absence of the sun at night was somewhat mitigated by reflected light from the moon. But, although the sun appeared on the horizon about the same time every day, the moon was more capricious and, even when full, provided no warmth. The moon was closer than the sun, and ancient man was able to study it in some detail. Before the time of Christ, the Greek astronomer Hipparchus determined the distance to the moon to be 240,000 miles. This was a remarkably accurate measurement that was improved upon very little by modern man with highly accurate instruments. Today, we know that the moon's orbit about earth is not a perfect circle, but is instead elliptical. The moon's perigee (the closest it get to earth) is 221,463 miles, and its apogee (the farthest it gets form earth) is 252,710 miles. Once the distance to the moon was known, it was very easy to determine the diameter of the moon to be 2160 miles. This is a measurement and calculation you can make.

What is the force that keeps earth in its orbit around the sun? Is this the same force that keeps the moon from escaping its earthbound orbit? If you answered "gravitational forces", you answered correctly. Each force is due to the attraction of one for the other, ie., the mass of the earth attracts the moon and the mass of the moon attracts the earth. The intersect of these two forces is about 215,000 miles form earth. You already know the mass of earth. The mass of the moon is 7.35×10^{22} kg and is 1/81.3 that of earth. Gravity on the moon is 5.33 ft/sec-sec or 1.6 m/sec-sec. Of what importance were these numbers to NASA when they were planning the Apollo 8 shot? (Apollo 8 was manned by three astronauts and circled the moon ten times before returning to earth on December 27, 1968.)

The rays of the sun always illuminate half of the moon's sphere and half of the earth. When the moon is between the earth and sun is the time of the new moon. Its dark side faces us and we cannot see the moon at that time. Two weeks later the moon is on the opposite side of earth which is the time of the full moon. In correctly determining the period of the moon, ie., the length of time of one complete revolution around earth, one must consider that earth also rotates as the moon moves through its period. This rotation of earth causes the moon to rise fifty minutes later each morning. This amounts to about six hours in a week. It takes the moon 27 1/3 days to complete its cycle. This is called the sidereal month. However, the earth and moon are also moving around the sun. This shift in position of the earth and moon combination with the sun, causes the moon to take two more days to complete its periodic motion. This is called the synodic month, which is more nearly our calendar month.

Progression Chart Showing Sidereal Month Advance to Synodic Month

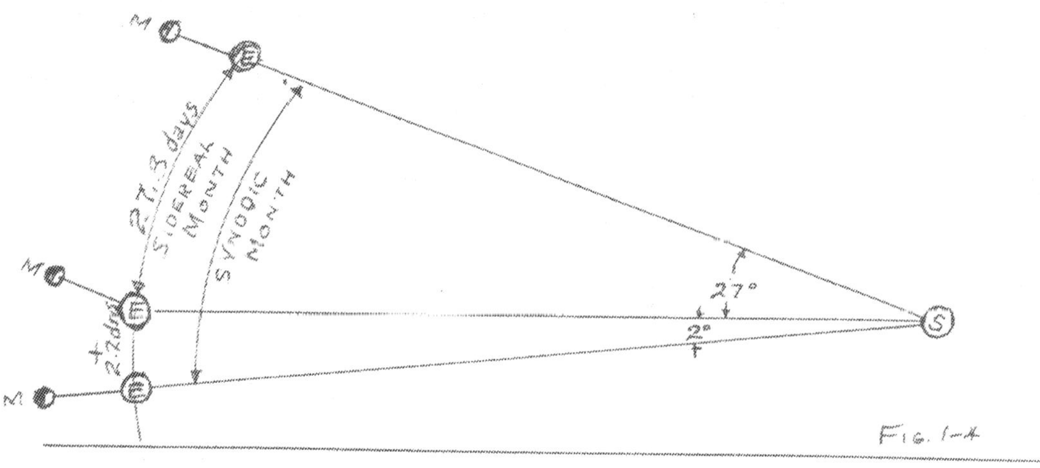

Searching the Face of the Moon

Man's inquisitive mind has led him on an endless search into the unknown, constantly looking for new lands and beyond new horizons. The surface of the moon has always brought forth his most imaginative speculations. Primitive man saw faces of people and animals in the dark and light areas of the moon. Galileo and others suggested that the dark spaces could be seas and the lighter areas solid land. These areas were given Latin names. The darker areas were called "maria" from the Latin "mare" for sea, and the areas that looked like bays were called "sinus", "oceanus" for ocean. With Galileo's invention and use of the telescope in 1609, and with subsequent improvement of the instrument, it became clear there were no seas or oceans on the moon and finally there was no water on its surface. Later on, it was evident there were large mountain ranges and rings, and many deep craters and rays. The mountains were given names such as Apennines, Caucasus, and Alps. By measuring the length of shadows, it was determined that the highest mountains in the Leibnitz Range are 26,000 to 30,000 feet high. When you consider that the moon is much smaller than Earth, these mountains would be 100,000 feet high on earth. (On a clear night, take your best binoculars or a small telescope and look carefully at the moon. You will be amazed at some of the detail you will be able to see.)

Better known than the mountain ranges are the craters and mountain rings. The Italian astronomer, Riccioli, named many of the lunar craters after well known astronomers of his time. The crater Copernicus is a series of interconnecting ridges surrounding a plane 40 miles in diameter, surrounded by walls which have peaks that are 56 miles apart. From its floor, mountains rise 300 meters, with slopes of 30 degrees. Clavius, one of the larges craters on the moon, has a diameter of 240 miles. Tycho, a somewhat smaller crater, can be seen with a small telescope at the time of full moon.

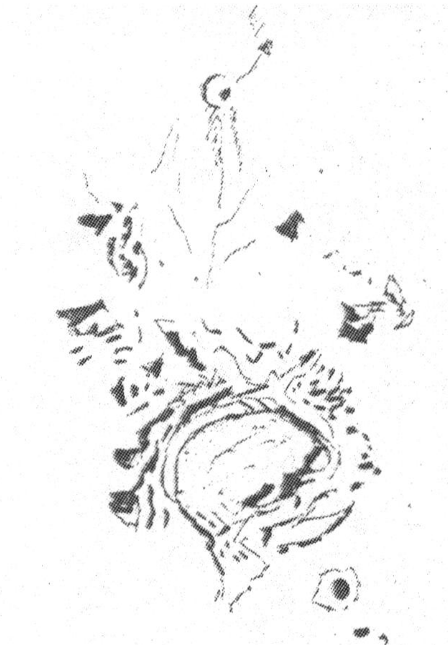

THE CRATER PLATO
Is known for its large, smooth bottom. It is located in the eastern portion of Mare Imbrium.

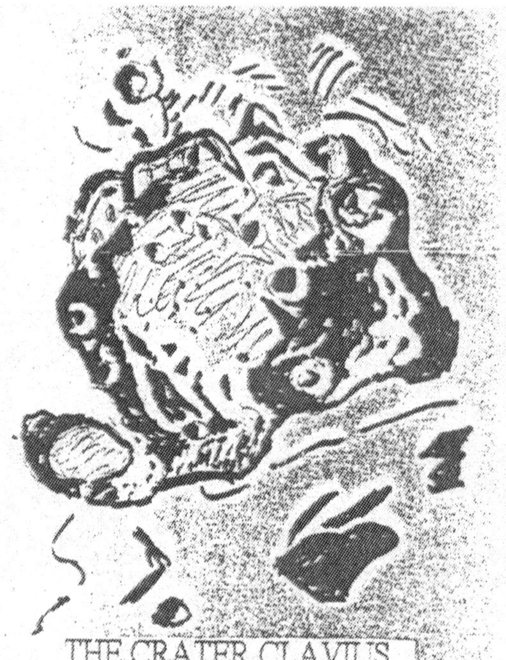

THE CRATER CLAVIUS
Is the largest crater on the observable side of the moon. It is 150 miles across.

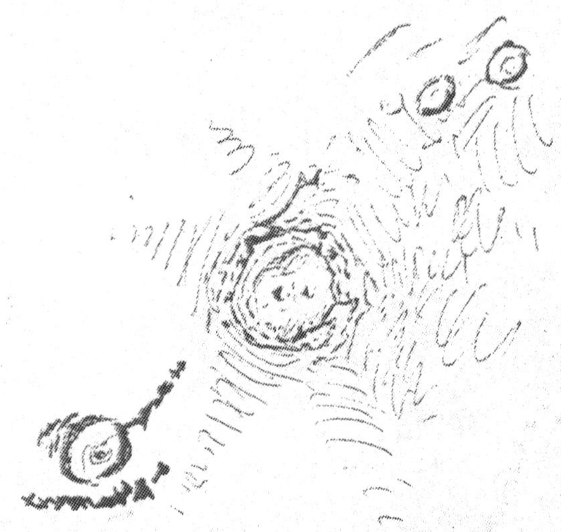

THE CRATER COPERNICUS
Is noted for its diverging rays. It is 50 miles across. On its floor are mountains which rise nearly 1000 feet.

THE CRATER TYCHO
Can be seen by a small telescope at full moon.

-9-

MAJOR MOON CRATERS

THE MOON'S EASTERN HEMISPHERE

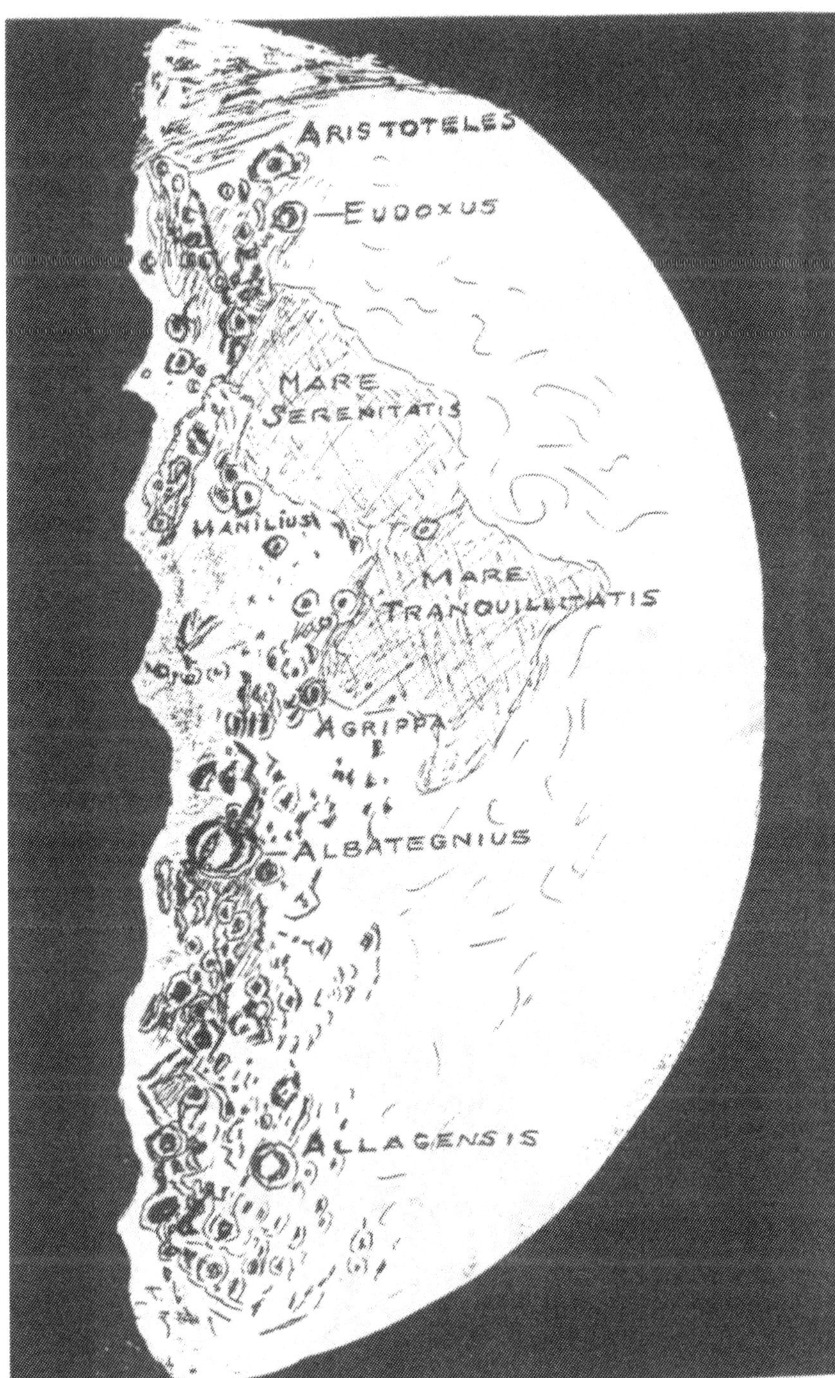

This drawing of the moon clearly shows the craters Aristoteles, Eudoxus, Manilus, Agrippa, Albetegnius and Allegensis. Mare Tranquillitatus and Serenitatis are clearly shown in the center of the moon's surface.

Measurement of Moon Distances

The picture of the eastern most part of the visible portion of the moon is as seen from earth. On a clear night, when the moon is full, you can see Mare Serenitatis and Mare Tranquillitatis from earth with the unaided eye. With good binoculars or a low powered telescope you may be able to see some of the other named features. You know the diameter of the moon to be 3475 kilometers. If you divide this number by 1.6 km/mi, you will recognize the number 2172 miles to be the English measurement. You are familiar with the fact that all maps are scaled down. If the vertical center line measurement of the pictures' diameter is 21 cm., what is the flat-line scale of this map using mi/cm (miles per centimeter)? Divide the true diameter of the moon by the measured length of the picture's diameter. This quotient is 103.4 mi/cm. (You will be working with mixed scale values, because you have a better concept of distance in miles than in kilometers and because the centimeter scale is measured in tenths of a centimeter rather than in English units.

It is important that you have some concept of the distances you will be measuring. For example, the driving distance from Los Angeles to New York City is approximately 3000 miles. This distance is about 1.38 times the diameter of the moon. The diameter of the earth is approximately 8000 miles or nearly four times that of the moon.

This will be your first of many assignments given in the written portion of the book. The first part is done for you as an example. You are to complete the chart below by making a chart of your own in which you write in the remaining names of "seas and "craters" and provide the correct measurements and calculations. Complete the "Flat distance" measurements first and then the "Curved distance" measurements next. Do Part B after you read the next full paragraph.

Exercise (1A) (Reference the Moon Chart on the previous page.)				
Locations	Part (A) Flat Surface Distances		Part (B) Curved Surface Distances	
On the Moon's Surface	Measure the Longest Dist. in (cm)	Convert (cm) into miles. Show Calculation.	Measure the Longest dist. in (cm)	Convert (cm) into miles. Show Calculation
Aristotel	0.7 cm	103.4 mi x 0.7cm = 72.38 miles		
Eudoxus				
Mare Seren-itatis	4.5 cm	103.4 mi x 4.5 cm = 465.3 miles		

(There are five additional names of craters and seas. Include all of these as part of your assigned work.)

THE CRATER TYCHO

THE CRATER TYCHO IS APPROXIMATELY
38 MILES ACROSS AND 14,800 FT. FROM
RIM TO BASIN

Did you stop to think that you were making measurements on a flat surface, but were working with a spherical surface? This makes your work inaccurate. Calculate the circumference of the moon and then calculate the new curved surface scale for the moon map. Don't forget that you are working with only one hemisphere of the moon. Was your answer 162.4 mi/cm? That number is considerably larger than 103.4 mi/cm. Also take note of the fact that one half the circumference of the moon is 1000 more miles than the distance across the United States. Now complete Part B of the chart, but keep your work until you have completed reading the next section on numerical error and percent of error.

Precision of measurement

Precision is a measure of the accuracy with which you work. For example, in a machine shop it is not unusual to hear machinists talk of working within tolerances of 1/10,000 of an inch or 1/100,000 of an inch. In order to do this, you need very precise instruments for your work. In physics, there are two ways to determine the accuracy of your measurement. One is the numerical error which is nothing more than subtracting two different measurements of the same object.

Numerical Error = Standard Value – Experimental Value = Num. Error

The standard value is the more accurate value and the experimental value is the less accurate value. Usually the value chosen is determined by the person making the calculation, and depends upon which tool was the more accurate tool used to make the measurement.

Percent of error is usually the best indication of the precision with which you have worked, because it uses the value of the numerical error and divides it by the standard value which gives a ratio of accuracy.

Percent of Error = (100) x (Numerical Error/(Standard Value)

So that you may better understand how this works, let the standard value be represented by the spherical scale constant = 162.4 mi/cm and the experimental value be the flat surface scale = 103.4 mi/cm.

% of Error = (100) x (162.4 mi/cm – 103.4 mi/cm)/(162.4 mi/cm) =
Percent of Error = 36.33% error

This means that the spherical constant is 36.33 % more accurate than the flat surface constant. In this case it was not the instrument used for measurement that was more accurate, but the method of making the calculation that had greater accuracy. You will be asked to make this calculation in the exercise at the end of this section.

Exercise 2

Self Test

(1) Calculate the average distance of the moon from earth using the apogee and perigee measurements.

(2) What part of the moon and the earth does the sun always illuminate?

(3) The rotation of the earth causes the moon to rise on the horizon how many minutes later each day?

(4) How long is the sidereal period of the moon? The synodic period?

(5) Make a diagram in which you label the positions of earth, moon, and sun for new moon and full moon. At what two positions in a circle of the above drawing would you place the moon for the half moon phase? (Place an (x) in your drawing to show the location of these positions.)

(6) Write the moon names that are counterparts of the following: (a) Sea; (b) Bays; (c) Oceans.

(7) Use the hemispheric map of the moon to measure the distance between the center of the craters Eudoxus and Agrippa in centimeters. Using the flat scale and the spherical scale, calculate this distance in miles. Calculate the percent of error between these two derived values. How close is your result to 36.33%?

Earth's Planetary Neighbors (The Solar System)

In addition to the sun and the moon, the Greeks had identified five other objects in the sky whose motion was different from that of the stars which they perceived as being fixed in the sky. These bodies known as planets. Ancient man perceived earth to be located at the center of this system and the stars. This system, the geocentric system (earth at center), was later to be named after astronomer, Claudius Ptolemaius. The system was called Ptolemaic System because of his written work around 130 AD.

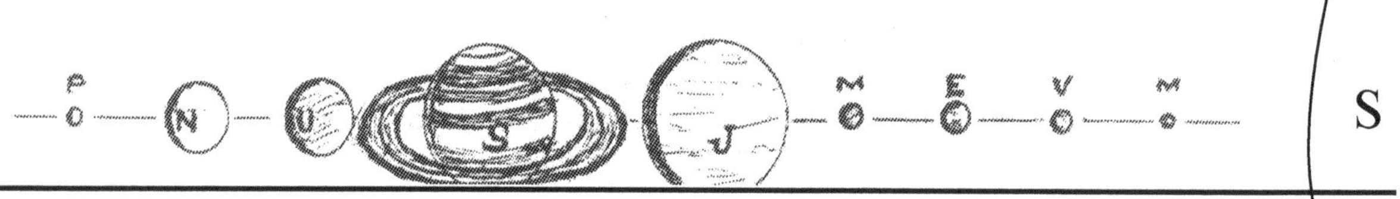

P M E V M S

Fig. 1-6

Pluto (Not yet discovered) Neptune Uranus Saturn Jupiter Mars Earth Venus Mercury

The Planets in Order and Their Relative Size

The Ptolemaic System, though questioned by many, was strongly supported by the church. Nicolas Copernicus, a Polish astronomer, was among many who began to believe the sun was the center of the Universe. Because he feared for his life, he did not publish his theories until 1543, on the day of his death. Having finally determined the true orientation of the sun relative to its planets, it now became possible to study the geometry of the motion of the planets about the sun. It was thought that the motion of the planets were the perfect circles because all of God's creation must be perfect. In 1609, Johannes Kepler was the first among his piers to advance an accurate model. He suggested that the orbits were not perfect circles, but were instead elliptical, ie., that the orbit was always an ellipse and the central body was always at one focus of the ellipse. With the publication of Kepler's Laws in 1619, and with the improvement of the refracting telescope, astronomers could now begin to make accurate determinations of the distances to all of the known planets.

The Average Distance of the Planets from the Sun		
Planets	Million Miles	Million Kilometers
Mercury	35.9	57.9
Venus	67.2	108.2
Earth	92.9	149.5
Mars	141.5	227.9
Jupiter	483.3	778.3
Saturn	886.1	1428.0

By 150 B.C., the Greeks, by triangulation, had accurately determined the distance to the moon. However, determining the distance to the sun was much more difficult. The Greek distance was five million miles which was very good for a time so early in history. Man was finally beginning to understand the immense size of the universe. Fifteen hundred years later, with the publication of Keplers three laws, scientists could now begin to calculate the relative distances of the planets from the sun. This permitted the formulation of an accurate model of the Solar System in which planetary orbits were laid out in correct proportion. The chart to the right is the result of one of the original studies.

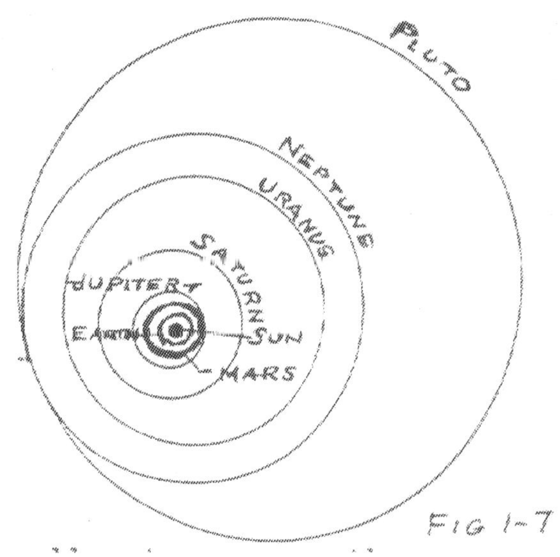

Astronomers had long used a method of parallax to measure the distance to the planets, but their measurements were inaccurate, because the unaided eye cannot perceive tiny objects accurately at stellar distances. With Galileo's reinvention of the telescope, his device became the principal instrument used by astronomers to make accurate observations of distant objects. In 1671 the astronomers Jean Richer and Giovanni Cassini made two simultaneous observations of the planet Mars. Richer made his observation from Cayenne in French Guiana. Cassini made his observation from his observatory in Paris, France. Cassini determined the distance from Earth to Mars to be 132 million miles and then used this measurement to recalculate the distance from earth to the sun to be 87 million miles. Using this distance to the sun, it then became possible to make a reasonably accurate determination of the distances of all of the known planets from the sun. (This is a good example of the important application of precision instruments used by scientists to make very accurate measurements.)

During the next three centuries the last three remaining planets were discovered. The planet Uranus was discovered by Wm. Hershel in 1781. In 1864, the French astronomer Jean Leverrier discovered the planet Neptune. The American astronomer Clyde Tombaugh discovered Pluto in 1930 while making observations at the observatory in Flagstaff, Arizona.

Average Distance from the Sun		
Planets	Million Miles	Million Kilometers
Uranus	1782	2872
Neptune	2792	4498
Pluto	3671	5910

A Modern Concept of the Solar System

In a very modern sense, the planets may be considered to have their combined origin in the sun. The immense gravitational reach of the sun is more easily understood in terms of its mass. The sun holds 99.86 percent of the solar system's mass. The diameter of the sun is 109 times greater than that of earth and its mass 333,000 times greater. It is the nearness of the planets to the sun that holds them so closely bound to it.

The early development of glass blowing, grinding, and polishing, made it possible to produce large numbers of increasingly powerful telescopes that could see farther and farther into space. The invention of spectroscopy then made it possible to investigate the chemical makeup of the sun and near-by planets. With the invention of radar, and the accidental discovery that it could be used to study radio emission from distant planets and stars, a vast amount of information about the composition of the planets and the sun became available to scientists. It was soon determined that the chemical composition of all of the planets was similar, but that variations in their size, density and mass caused some chemical differences. For example, the four inner planets, the "terrestrial planets" are composed mostly of rock material. The "Jovian planets", Jupiter, Saturn, Uranus and Neptune, are composed of the lighter elements which are mostly hydrogen and helium. Pluto does not fit either of these groups because it is composed mostly of ice.

The chart below will provide most of the current statistical data about the planets that you may need for reference in this course:

Planetary Data								
Data	Mercury	Venus	Earth	Mars	Jupiter	Saturn	Uranus	Neptune
Av. Dist from sun A.U.	.387	.723	1.00	1.52	5.2	9.54	19.18	30.06
Mean Dia. E=1	.38	.96	1.00	.53	11.19	9.47	3.73	3.49
Mass E=1	.055	.815	1.00	.11	318	95.2	14.6	17.2
Rotation Period	58.6d	-243	23.9h	24.6h	9.8h	102.h	-12.3h	16h
Surface Gravity E=1	.37	.88	1.00	.38	2.64	1.15	1.17	1.18
Known Satellites	0	0	1	2	13	10	5	2
Data for Pluto: 39.44; .25; .06; 6.4d; .2; 0.								
1 A.U. (Astronomical Unit) = 150,000,000 kilometers)								
The negative sign in front of the period of rotation indicates reverse rotation.								

Exercise 3

Self Test

1. Name the five planets known to Greek astronomers.
2. What name was given to the "Copernicus" System?
3. Discuss Kepler's principal premise.
4. According to Kepler, what body was at the focus of any planetary orbit?
5. What method did early astronomers use to determine the distance to the moon?
6. Name the type of telescope invented by Galileo. (a) Did this telescope use a mirror?
7. Cassini determined the distance to the sun to be 87 million miles. Use the value given in the accepted value for this distance to calculate his percent of error.

Use the Planetary Data Chart on the previous page as your reference source in making all calculations for the problems below. Remember that earth data is the basis for all planetary answers.

8. What distance is Saturn from the sun?
9. What is the mean diameter of Jupiter?
10. What is gravity on Mercury (English units only)?
11. What is the rotational period of Mercury in hours?
12. What is the diameter of Mercury in metric and English units?

Donald C. Lundy

PART II – Physics

Donald C. Lundy

Chapter II

Subjects Basic to the Physics of Space Travel

Scientific Calculators

Now that you are familiar with facts related to our Solar System, you can turn your mind to the mathematics that will determine your ability to travel within the bounds of our Solar System. By this time, you should have purchased a hand-held, scientific calculator. The Radio Shack calculator below was chosen because you can easily see the functions that you will need to use in making scientific calculations. Texas Instruments (TI), Sharp and others are equally as good as the 'LCD Scientific' calculator shown below. We are mostly interested in the upper two rows of functions, ie., log 10, In e, sin, cos, tan, power and root functions. This calculator is a 2 level type, the second level achieved by pressing the (INV) key. Newer calculators have three levels of functions. But, before you get underway with your calculator, you need to review 'powers of ten notation.'

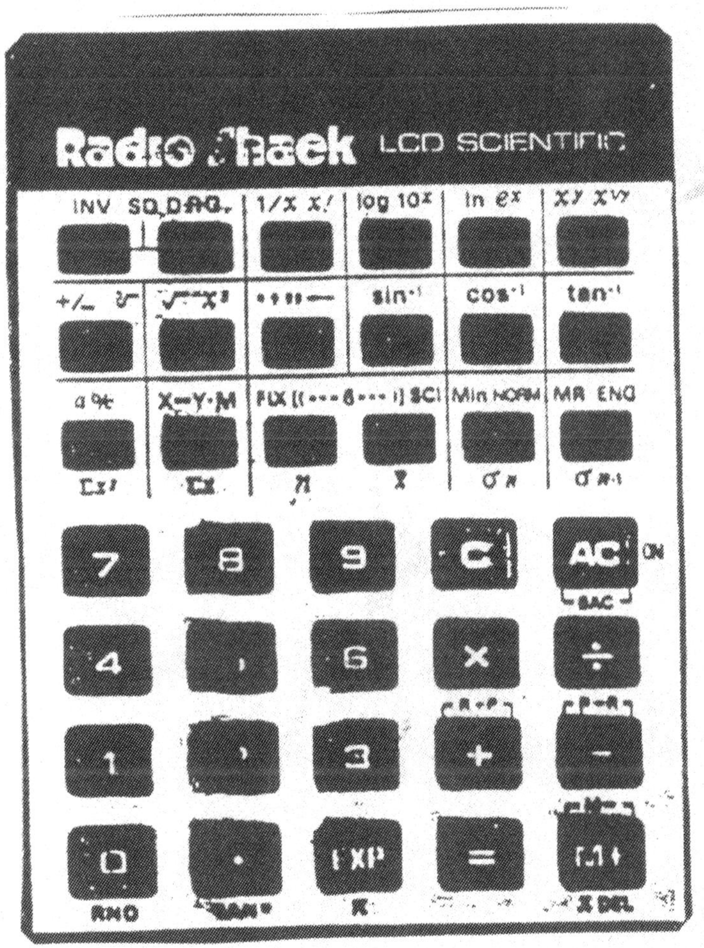

Powers of Ten Notation

You were introduced to some very large numbers in the first chapter. For this reason, you should understand the need for simplifying large numbers. What follows is a series of self tests preceded by a brief explanation of the exercise. Rule 1: Numbers may be expressed in a standard form by rewriting them so that only one place is to the left of the decimal point and expressing the number as a power of ten—the power representing the number of places the decimal has been moved. Two examples follow:

(a) $24{,}500 = 2.45 \times 10^4$ (b) $.000245 = 2.45 \times 10^{-4}$

At times it is important to reverse the above process in order to express standard notation in the form of a decimal numeral. Rule 2: Using the power of ten, count from the decimal to the right for positive powers and to the left for negative powers and write the number in normal form. Two examples follow:

(a) $7.31 \times 10^8 = 731000000.$ (b) $4.13 \times 10^{-8} = .0000000413$

As numbers become more complex in interaction it becomes necessary to combine the product and quotients of several powers of ten. Rule 3: When two or more powers of ten are to be multiplied, write the number ten and affix the power as the algebraic sum of the combination of powers. When dividing, write the number ten and affix the power resulting from the subtraction of the power in the denominator from that of the numerator. Two examples follow:

(a) $10^5 \times 10^7 = 10^{5+7} = 10^{12}$ (b) $10^9/10^5 = 10^{9-5} = 10^4$

Rule 4: In complex problems involving powers of ten, combine all of the powers in the numerator writing a single power of ten and do the same in the denominator. Then combine the powers as in Example (b) above.

$(10^6 \times 10^{-2} \times 10^3)/(10^{-1} \times 10^4) = 10^7/10^3 = 10^4$

Rule 5: When both decimal numerals and powers of ten are involved in complex situations, combine all of the powers of ten into a single power of ten and use your calculator to derive a common answers for all of the decimal numerals. Write the answer as the product of a decimal numeral and a single power of ten.

(a) $4.511 \times 10^3 \times 9.7 \times 10^7 \times 3.22 \times 10^{-2} = 140.9 \times 10^8$

Rule 6: When raising a power to a power, given the base is ten, write the common base, ie., ten and affix the product of the powers. Two examples follow:

(a) $(10^6)^5 = 10^{6x5} = 10^{30}$ (b) $(10^{-6})^{1/2} = 10^{-6x1/2} = 10^{-3}$

Exercise 4

Self Test

See how well you can do with the following problems. Use your calculator where appropriate. You will find all answers in the back of the book.

1. Write the following in standard notation:
   (a) 237,000; (b) .000957

2. Write the following as decimal numeral:
   (a) 9.72×10^{-3}; (b) 6.341×10^{5}

3. Write as a single power of ten:
   (a) $10^2 \times 10^0 \times 10^5 \times 10^{-3}$; (b) $10^6/(10^5 \times 10^4)$; (c) $(10^3)^4$

4. Write the answer in standard notation:
   (a) $9.81 \times 10^{-5} \times 1.3 \times 10^4 \times 7.9 \times 10^6$; (b) $(4.821 \times 10^{-1} \times 3.27 \times 10^5)/(6.8 \times 10^6)$

5. Write the following problem in standard notation. You need not solve the problem.
   (a) $\frac{17.045}{9.72} \times \frac{973}{5937}$

You can hardly wait to put your scientific calculator to use after standard notation. The following exercises will give you a chance to become familiar with your calculator. If you find some of them perplexing, check in the instructional booklet that came with your calculator. When you finish this exercise successfully, you are ready to move into some of the other calculations that will really put your expertise to effective use. Use the answers provided by your calculator as correct answers for the following:

1. Square 3.92
2. Find the square root of 7.38
3. Find the reciprocal of 2.58
4. Cube 2.71
5. Find the cube root of 738
6. Raise 8.2 to the 2.5 power
7. Find the common log of 27
8. Find the natural log of 27
9. Given the angles 38°, 131°, 280° find their sin, cos, and tan functions.
10. Given the function of an angle to Be .562, find the sin, cos, and tan Angles represented by this function.
11. $\left(\frac{.0073x623.8}{\sqrt{32.4}}\right)^2$
12. $\left(\sqrt{8.33}\right)^8 \times \left(\sqrt[3]{800}\right)$
13. $\left(\frac{9.17}{3.96}\right)^2 \times \left(\frac{47.38}{962.3}\right)^3$
14. $\left(\frac{\sqrt{328.5}}{\sqrt{.68}}\right)^{\frac{5}{3}}$

<u>Graphing Equations</u>

Graphing provides an easy, pictorial way to look at formulae and equations. Consequently, graphing makes possible a simplified way to view the laws of science and mathematics. Equations have been devised for all of the basic geometric architectural designs and for the paths followed by the planets about the sun and the moons about their planets. All such paths are elliptical. There are equations that trace the take-off and landing trajectories of rockets. Some of these paths are drawn in the chart below, to the right. There are straight line graphs and curved graphs. The interpretation of these graphs make for a more complete understanding of related technical information about space vehicles and their thrust capabilities with specific types of fuels for example. Graphing is easy and fun. You need only two perpendicular, intersecting lines called axes. Each line is labeled and numbered for meaning. An example of a straight line graph with data is drawn below.

Example: Make a graph of the relationship between (F) and (a), (Force and acceleration), in the equation F=(m)(a). This is Newton's Second Law of Motion. In this equation mass Equals 1,ie., m=mass=1.

F = (m)(a)		
F(pounds)	a = F/m	a(ft/sec
0	0	0
1	1	1
2	2/1	2
3	3/1	3
4	4/1	4
5	5/1	5
6	6/1	6
7	7/1	7
8	8/1	8
9	9/1	9
10	10/1	10

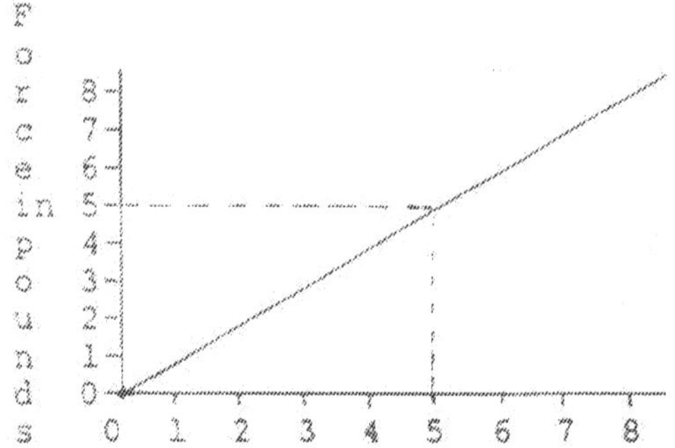

Acceleration in (ft/sec)

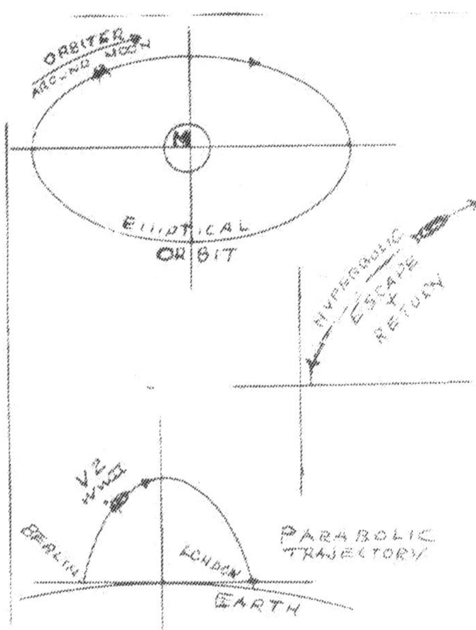

Above are drawn several of the types of orbits, entry and launch to orbit or escape patterns, of <u>rockets and orbital vehicles.</u>

1. Given the formula (C= (pi) (d)) for the circumference of a circle. Arrange a chart with two columns labeled C for the circumference and D for the diameter in feet. Represent D with the whole numbers 1 through 10 and calculate C for each value of D. Make up a graph in which you represent all D values on the horizontal axis and all derived values of C on the vertical axis.
2. Given the equation $y=-x^2 + 10x$, make up a data chart in which the x values range from 0 to 10 and calculate the derived values for y. Graph x values on the horizontal axis and y values on the vertical axis. You will recognize this graph as that of the parabolic path of any object which is shot into the air and falls back to earth.

Trigonometric Functions

In this part of your study, you will be reviewing lines, angles and some other geometric figures. The only new concepts will have to do with vector quantities which will begin your study of the sections of physics which will be needed to do all space calculations. You will need an English-metric scaled ruler and a protractor to measure angles. Without these tools, you will not be able to complete your work.

To the right is the classic right triangle of Pythagoras. The sides are labeled (a), (b), (c), and the angles opposite each side are labeled (A), (B), (C). In trigonometry, there are three simple but important ratios that relate angles and sides. These are as follows:

To the right is the classic right triangle of Pythagoras. The sides are labeled (a), (b), (c), and the angles opposite each side are labeled (A), (B), (C). In trigonometry, there are three simple but important ratios that relate angles and sides. These are as follows:

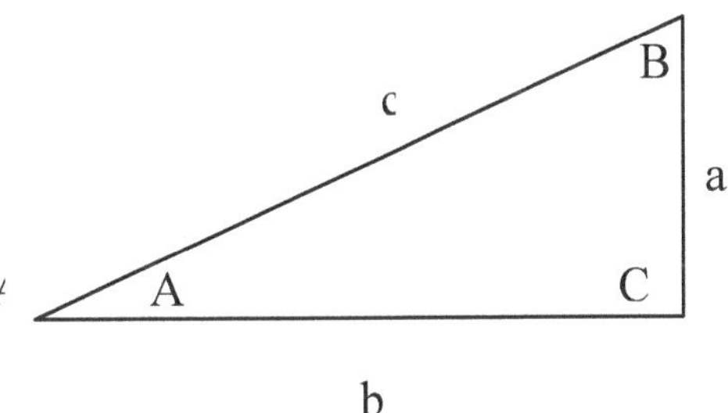

$$\text{Sin A} = \frac{\text{Side opposite}}{\text{hypotenuse}} = \frac{a}{c}$$

$$\text{Cos A} = \frac{\text{Side adjacent}}{\text{hypotenuse}} = \frac{b}{c}$$

$$\text{Tan A} = \frac{\text{Side opposite}}{\text{adjacent}} = \frac{a}{b}$$

You can see that the above ratios can be used with any of three angles as long as the adjacent side are perpendicular to each other.

Exercise (8)
Self Test

Measure the lengths of the sides of the above triangle in centimeters and their opposite angles. Make up a chart of sides and angles. Include in your chart the sin, cos, and tan ratios for angle (A). This is a simple exercise that will acquaint you with your ruler and protractor.

Right Triangles Continued-The Pythagorean Theorm

There are times when you do not know the angles of a right triangle, but you know the length of two of the sides. Such a triangle is drawn to the right. Pythagoris stated that if two adjacent sides are perpendicular to each other, then the square of the hypotenuse is equal to the sum of the squares of the other two sides.

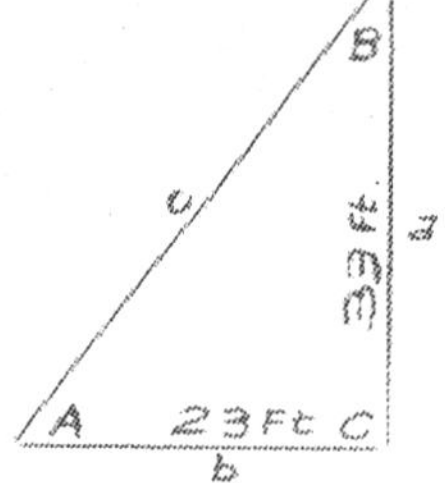

$C^2 = a^2 + b^2$ or $c = \sqrt{a^2 + b^2}$

Using the above equations to solve for the third side of the drawn triangle, it follows that:

$$c = \sqrt{(33ft)^2 + (23ft)^2} = \sqrt{1618ft} = 40.22 \text{ ft}$$

Use the tan function of angle (B) to calculate that angle and then check this result using the sin function of angle (B).

Tan B = 23 ft / 33 ft = .697 or B = 34.88°
Sin B = 23 ft / 40.22 ft = .572 or B = 34.88°

Other than the right triangle, there are two other kinds of triangles which are isosceles and equilateral triangles. The functions used on right triangles do not work on these triangles, but many times these triangles can be subdivided into two right triangles and solved for their sides and angles in that manner.

Application of Trigonometric Functions to Vector Problems

There are two types of quantities in physics, ie., vector quantities and scalar quantities. Vector quantities are those which have both magnitude and direction. Scalar quantities have only magnitude. In the right triangle at the top of the page, the lengths 23 ft and 33 ft are scalar quantities. However, if each is described with direction, ie., 23 ft West and 33 ft North, both measurements become vector quantities. Each can now be drawn to scale and represented by an arrow which shows direction. Distance, velocity, weight and force are examples of vector quantities.

Drawn to the right is a reference circle that will show how to draw angles form 0° up to 360° and how to reference these angles along vertical and horizontal axes. As most protractors form angles on a half circular plane, yours will work best when placed above or below the horizontal axis.

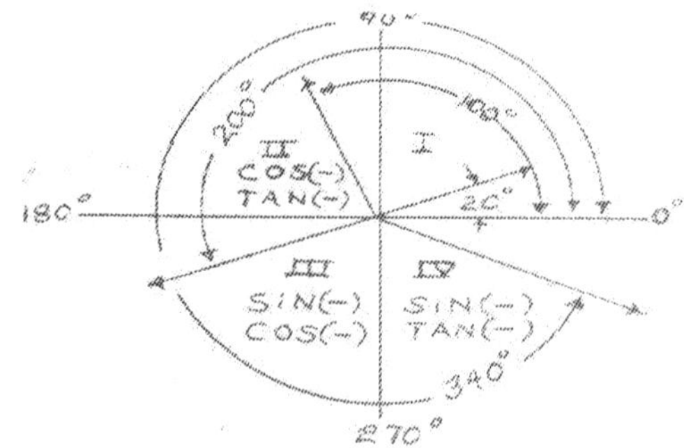

The Reference Circle of Quadrants and Function Signs

Looking at the 'reference circle,' you can see that it has four quadrants. Using compass, map designations, you may think of 0° as East, 90° as North, 180° as West and 270° as South. This would be normal map orientation. Now use your calculator to check the sign designation of angles in each quadrant. Find the sin, cos and tan designations of the angles shown in the circle. You will find that all functions are (+) in the first quadrant, but that cos and tan are both negative in the second quadrant, ie., the cos 100° is -.174 and the tan 100° is –5.67. Similarly, signs of functions are shown in each of the other quadrants.

Resultant and Equilibrant Quantities

The answer that you obtain by adding two or more vectors geometrically or algebraically is called the resultant (R). Where practical, you may use graphical or geometric solutions of these problems. In solving this type of problem, you first set up a scale that permits you to make an accurate drawing of the physical situation on paper. In situation (a), you walk 10 meters directly to the west and then reverse your direction and walk 15 meters in the opposite direction. Choose a scale such as 1 cm = 1 m and draw an arrow 10 cm long pointing to the west. Then draw another arrow 15 cm long pointing to the east. The resultant value (R) = 5 meters pointing eastward. In this case it is obvious that the numbers can be added algebraically to get an accurate solution. If you make your drawings large enough and accurately, your results are acceptable.

The equilibrant is defined as that vector which is equal in magnitude to the resultant but acts 180° opposite to the resultant. The equilibrant is drawn and labeled as an opposite extension of the 18N resultant shown in Figure (b) below. Think of the equilibrant as that force which acts to produce equilibrium in a given physical situation. In arm wrestling, when you apply enough force so that neither arm is capable of moving, you have provided the equilibrant force.

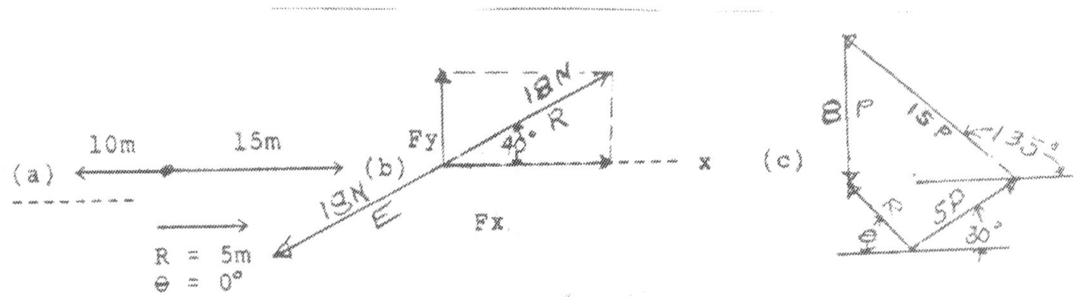

Exercise 7

Self Test

1. (a) Write the tan 77°; (b) the cos 177°; (c) the tan 233°.

2. (a) In the previous example, ref. Fig (b), calculate the Fy component and (b) the Fx component of force.

3. In Fig © on the previous page, calculate: (a) the resultant (R); (b) the angle 0, and (c) name the equilibrant. (The drawing is not to scale.)

4. Given a resultant force of 42 p acting at an angle of 142°; (a) calculate the Fx and Fy force components; (b) write the direction and magnitude of the equilibrant.

5. Given the triangle drawn to the right: (a) use the Pythagorian equation to calculate the third side and (b) calculate the acute angles.

6. Given a right triangle and sides a = 15 ft and c = 31 ft, find side (b) using the Pythagorian equation.

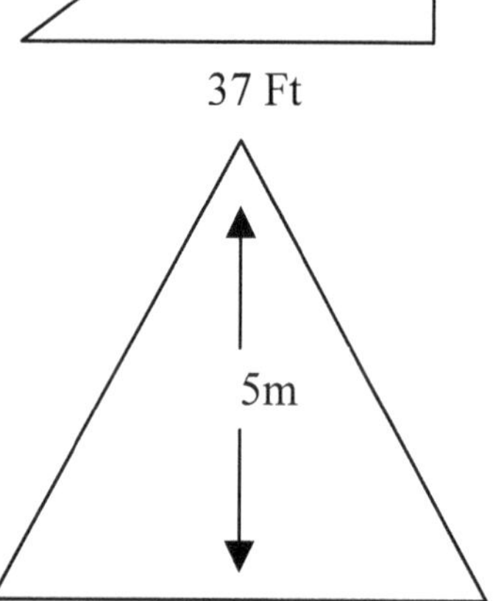

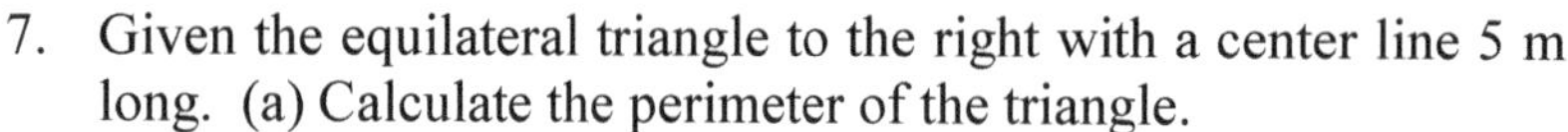

7. Given the equilateral triangle to the right with a center line 5 m long. (a) Calculate the perimeter of the triangle.

8. While looking for a lost object in a field, you walk 20 paces S., then 10 paces SW., then 40 paces NW., and finally 15 paces E. How far are you from your starting point and what is the angle from this point back to where you finished? (Hint solve this problem with a similar method used in Example (c).

The Component Method of Solving Vector Problems

In the component method of solving vector problems, a given set of vectors are sub-divided into their respective x and y components by using either the parallelogram method of solution or multiplying each by the sine or cosine of its respective angle. Then all x and y components are algebraically added, this resulting in a single x component and y component. These two forces are then combined into a single vector.

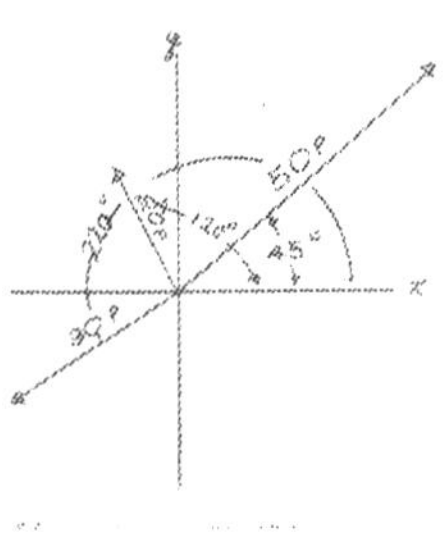

Three forces Fa, Fb and Fc are drawn in three different quadrants with magnitudes and directions as shown. The same forces are drawn separately below, where they are broken down into x and y components. They are designated again in the chart where their respective x and y components are shown calculated with results.

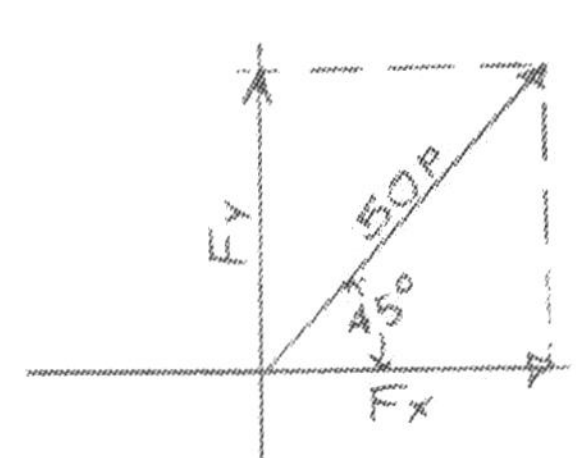

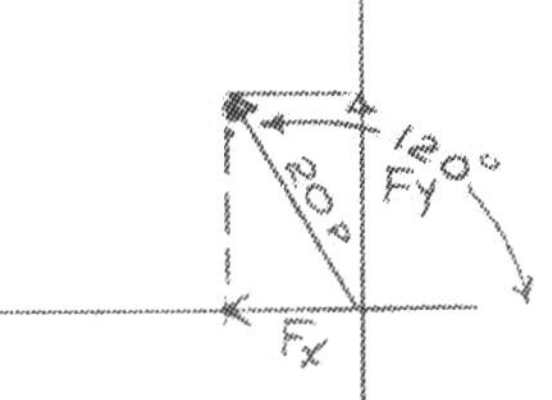

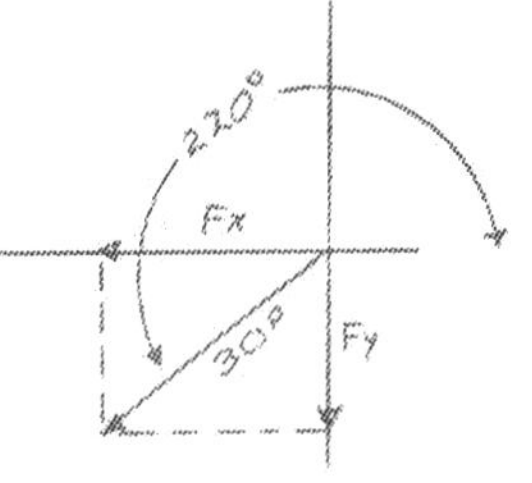

Component Chart of Vector Forces			
Calculations			
Forces	Angle	Fx Calculations	Fy Calculations
Fa 50 p	45°	(50p) (Cos 45°) = 35.35p	(50p) (Sin 45°) = 35.35p
Fb 20 p	120°	(20p) (Cos 120°) = -10p	(20p) (Sin 120°) = 17.3p
Fc 30 p	220°	(30p) (Cos 220°) = -23p	(30p) (Sin 220°) = -19.3p
SUM OF THE COMPONENTS = +2.35p			+33.35 P

$$R = \sqrt{(2.35p)^2 + (33.35p)^2} = \sqrt{1117.7p} = 33.43p$$

Tan 0 = (33.35P)/(2.35P) = 14.19 0 = 86°

Chapter III

The Dynamics of Motion

Space is dynamic in that all matter moves relative to all other matter. Motion here on planet earth is fundamental to life. For this reason, it is important to learn as much about the physics of motion as you can.

Velocity, Time, Distance Relationships

If you are riding in a car, you can tell how fast you are going by the roar of the engine or by looking outside at stationary objects as you pass by. You may say that you are going slower or faster or that you are moving at high speed. At best, these expressions mean relatively little and are not precise enough to have any real meaning in the physical world. If you look at the speedometer, you have a much better concept of your speed. In physics, the word speed is seldom used, instead we define motion in terms of velocity, distance, and time. Distance (D) is defined as displacement or change of position. The units of distance most familiar to you are miles, feet and inches, but such metric units as centimeters (cm), meters (m), and kilometers (km) are also very important. Time is measured in years, days, hours, minutes and seconds – that part is easy. You are familiar with the units of velocity which are designated on your speedometer. These units are mainly mph (miles per hour) in this country, but in most other countries are mainly mph (miles per hour) in this country, but in most other countries are designated as kilometers per hour (kph). In physics the units of velocity are written mi/hr, km/hr, ft/sec, m/sec and cm/sec. Velocity is defined as the distance traveled per unit of time.

Velocity = Distance/Time	V=D/T	D=(V) (T)

V, T, D, Relationships		
Distance	Time	Velocity
1 mi = 5280 ft 1 mi = 1.6 km 1 km = 1000 m 1 m = 100 cm 1 in = 2.54 cm	1 hr = 60 min = 3600s 1 min = 60 sec 1 yr = 365 days	1 mi/hr x 5280 ft/mi/3600 s/hr 1.467 ft/sec 1 km/hr x 1000 m/km/3600 s/hr .278 m/sec 1 ft/s x (12 in/ft) x (2.54 cm/in) = 30.48 cm/sec

The above chart is a reference chart which you are to use as an aid in making English-metric conversions.

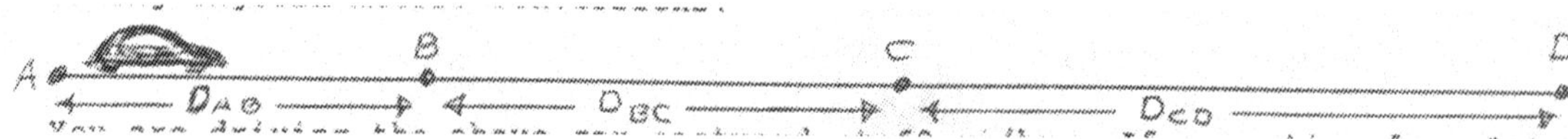

You are driving the above car eastward at 60 mi/hr. If your time from A to B is 0.2 hrs and from B to C is 0.3 hrs, and from C to D is 0.5 hrs. (a) Calculate the distance AB, BC, and CD in miles. (b) What total distance did you travel in km? (c) What is your average velocity in ft/sec?

D = VT then D(AB) = (60mi/hr)(.2 hrs) = 12 mi; D(BC) =18mi; D(CD) =30mi
D(total)=12mi + 18mi +30mi = 60mi. and (60mi)(1.6km/mi)=96km.

Did you calculate the V (average velocity) to be 88 ft/sec? Good work. You already knew that uniform velocity is the same as average velocity.

Exercise 8

Self Test

1. Convert 1000 yards to (a) feet, (b) miles, (c) kilometers, and (d) meters.
2. Make the following unit conversions: (a) 20 mi/hr to ft/sec, (b) 44 ft/sec to mi/hr, (c) 80 km/hr to mi/hr.
3. A rocket travels 4000 miles in 15 minutes. Calculate its velocity in (a) mi/hr, (b) ft/sec, and (c) km/hr.
4. A Delta Winged bomber flies at 2000 mi/hr for 12 hrs. (a) How far does it go in miles, (b) in kilometers?
5. If the moon's apogee is 252,710 mi, what must be the velocity of a rocket launched from earth to reach the moon in three days?

Accelerating Objects

Your experience tells you there is another kind of velocity. You may think of this as variable (changing) velocity. (Vi) is the symbol used to represent this velocity. The velocity is derived from uniform acceleration or deceleration. This is what you see on your speedometer when you are accelerating from a light. The faster you accelerate the faster the pointer changes position or the faster the change in digital readout. The formula for acceleration is as follows:

Acceleration = (instant velocity)/time = (Vi)/t

The units of acceleration are defined in terms of velocity and time.

$$a = \frac{v}{t} = \frac{(ft/\sec)}{\sec} = \frac{(cm)/\sec}{\sec} = \frac{(m)/\sec}{\sec}$$

It is customary to write: ft/sec^2; cm/sec^2; m/sec^2

Example (1) While driving your car, you stop for a signal light. When the light changes, you accelerate to 88 ft/sec in ten seconds. (a) Calculate your acceleration. Was your answer 8.8 ft/sec^2?

(b) Next calculate your velocity (v) at the end of 2 sec., 5 sec., and 7 sec.

V = a x t = 8.8 ft/sec^2 x 2 sec. = 17.6 ft/sec

(c) Calculate your average velocity fro the trip.

V = (v_o = v_i)/2 = (p ft/sec + 88 ft/sec)/2 = 44 f/s

Accelerated Distance

Every human being has a profound respect for gravity. Already, you are more familiar with its affinity for all matter than most people because of your short encounter with astronomy. The formula that gives the relationship between gravity and time during free fall is

Height or distance = (1/2) x (gravity) x (time)^2 = H = (1/2) (g) (t)^2

Example (2) On a warm day you climb upon a diving platform and let yourself fall into the water below. If the time of fall is 1.5 sec, (a) how high up were you in meters? (b) in feet?

H = .5 (9.8 m/s^2) (1.5 s)^2 = 11.03 meters

(You calculate this distance in feet.) Was this very high? (c) With what velocity in ft/sec did you impact the water?

V = gt = (32 ft/sec^2) (1.5 sec) = 48 ft/sec

Example (3) If you shoot an arrow vertically into the air with an initial velocity of 180 ft/sec, (a) How high will it go and (b) how long before it hits the ground? Move before you calculate (t)!

First calculate the time up: Tup = v/g = (180 ft/sec)/(32 ft/sec^2)

Tup = 5.63 seconds Total time = 11.25 seconds

The final acceleration formula has great application to rocket propulsion, as rockets are always being accelerated and decelerated.

$$\text{Acceleration} = \frac{change\_in\_velocity}{change\_in\_time} = \frac{(\Delta v)}{(\Delta t)} = \frac{(v_f)-(v_o)}{(\Delta t)}$$

The symbol delta (Δ) means change, ie., change in velocity and change in time.

Example (4) Assume that you are riding on the cone of an Atlas rocket which is well on its way to the moon. If your current velocity 18000 mi/hr and you are accelerated to 21000 mi/hr in 60 seconds, calculate your acceleration in mi/hr/sec.

$V_o = 18{,}000 \frac{mi}{hr}$ $V_f = 21{,}000 \frac{mi}{hr}$

a = (V_f – V_o)/(Δt) = (21,000 mi/hr – 18,000 mi/hr)/(60sec) = a = 50 mi/hr/sec. Change this to ft/sec/sec.

What do the acceleration units tell you about your motion? The units relate that every second the velocity increases by 50 mi/hr, ie., that your total velocity form the time of fire is 18,050 mi/hr end of first second of ignition; 18,100 mi/hr end of second second; 18,150 mi/hr end of third second, ...21,000 mi/hr at burnout. But the units are not acceptable units for acceleration. The correct units are derived as follows:

$$\frac{Mi}{hr-\sec} x \frac{ft/mi}{\sec/hr} = \frac{ft}{\sec-\sec}$$

50 (mi/hr-s) x 5,280 (ft/mi)/3600 (sec/hr) = 73.33 (ft/sec^2)

All of the previous discussion has been about objects accelerated from a rest position. How do you handle the mathematics of objects which are accelerated after they are set into motion? In order to solve these problems you first find the distance the object will have traveled do to its velocity at that point where the acceleration begins and then add to this distance the distance the object will travel during its period of acceleration. A handy equation is as follows:

$$S(tot) = (Vi)t + \frac{1}{2}(a)(t)^2$$

Exercise 9

Self Test

(1) When the signal light changes from red to green, you accelerate your car at the rate of 15 ft/sec-sec for 15 seconds. (a) What is your velocity at the end of 15 sec? (b) How far did you go during this period of time?

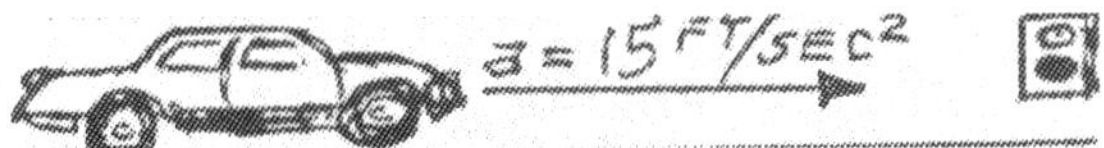

(2) Starting from a rest position, a skier accelerates uniformly for 400 yards. If he achieves a final velocity of 60 mi/hr, (a) what was his acceleration in ft/sec-sec and (b) how long did it take him to travel this distance?

(3) Starting from a rest position, you blade skate down a hill accelerating at 12 ft/sec-sec. What is your instantaneous velocity at the end of (a) the first second, (b) the second second, (c) the third second, (d) the sixth second, and (e) the ninth second?

(4) You are driving your car at the legal speed in a 45 mi/hr zone and then accelerate your car at the rate of 11 ft/sec-sec for 10 sec. (You must convert all units to ft/sec and to ft/sec-sec in order to solve this problem.) What is your final velocity at the end of ten seconds (a) in ft/sec? (b) in mi/hr?. (c) How far did your car go in feet due to its acceleration? (d) What total distance in miles did it travel during this period of time?

<u>Trajectory</u>

You now have most of the basic equations that are needed to predict the motion of an object which is launched from the ground. This part of the study will involve objects that can be thrown and shot from almost any kind of projecting or lofting device. For theoretical purposes, they must be small, smooth and of heavy density, so that air resistance has little or no affect upon them.

Example (1)

The first illustration is that of an arrow shot from a bow so that the arrow is launched parallel to the ground with a velocity of 100 ft/sec. Assume, at the moment the arrow is launched, a ball bearing is dropped form the same elevation six feet above the ground. The target is a can placed at ground level 80 feet away.

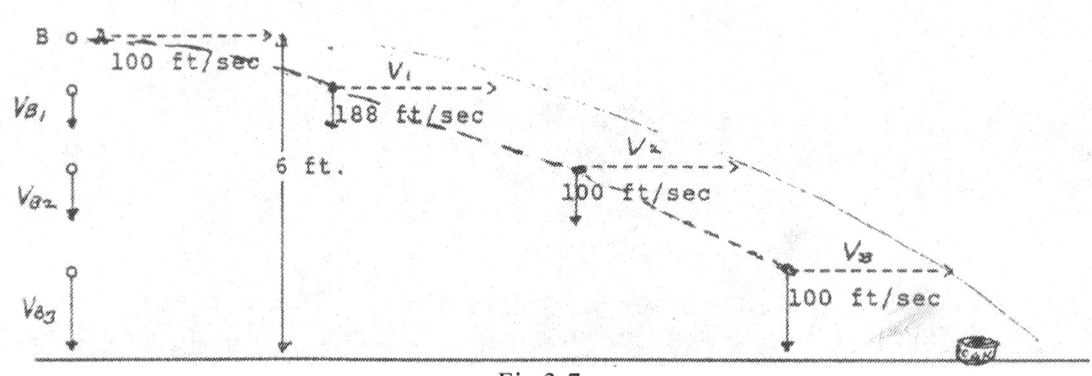

Fig 3-7

In the above drawing, vectors are used to help explain the vertical and horizontal action of the arrow and the ball bearing. Notice that the vertical component of velocity has nothing to do with the horizontal velocity which does not change. Why is the horizontal velocity unchanged? Is the horizontal component of velocity to be considered a uniform, constant velocity? Make the following calculation:

(a) How long before the bearing hits the ground?
(b) How long before the arrow hits the ground?
(c) How far will the arrow fly in this length of time?
(d) Is it possible for the arrow to strike the target?
(e) What must be the initial velocity of the arrow in order for it to hit the can?

(a) Solve the free fall formula for (t).	(b) The time is the same.	(c) How far will the ball go?
$t = \sqrt{(2H/(g)}$	V = d/t = (80 ft)/.612	D = v x t
T = .612 sec.	V = 130.72 ft/sec	D = 61.2 ft.

Parabolic Trajectory

Example (2)

That was a fairly easy trajectory problem. You are well aware that the arrow could have been shot up into the air. Next use the same conditions as in the previous problem except this time the arrow is shot into the air at an angle of 30°, and the target is now set six ft. above the ground. (a) How high will the arrow go before starting down? (b) How long will the arrow be in the air? (c) What will be the range of the arrow (how far does it go)? (d) Will the arrow reach the target while in flight? Begin by analyzing the path of flight, using vectors to understand what is happening to the arrow while in the air.

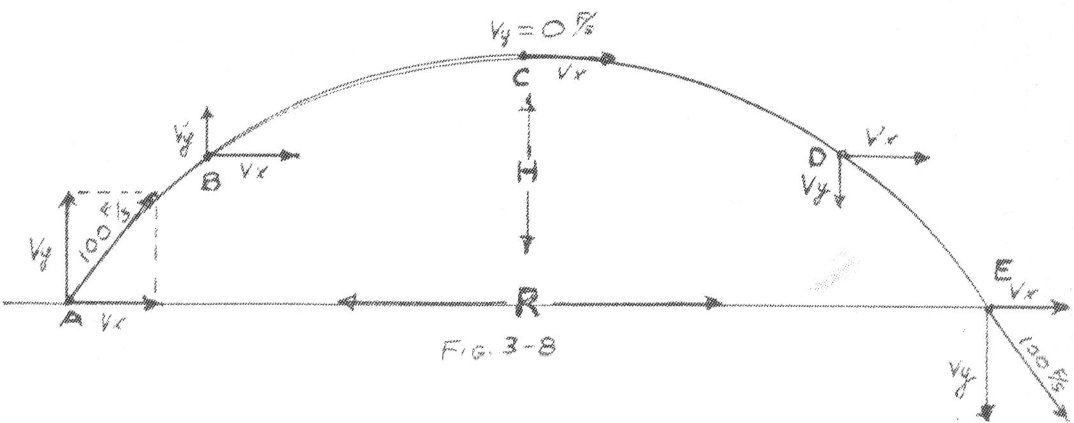

FIG. 3-8

Begin by analyzing the trajectory path. Notice that vector drawings (A) and (E) are respectively at the beginning and end of the course of travel of the arrow. Also, notice that drawings (B) and (D) are comparable drawings except that the (B) drawing represents the arrow in upward flight and the (D) drawing shows the arrow in downward flight. (C) is positioned at the top of the arc, at the center of the range distance. The first step in solving this problem is to make a vector diagram in which you show the arrow positioned at 30°. This vector represents the initial velocity of the arrow. Enclose the arrow in a parallelogram in which the vertical direction is Vy, and the horizontal direction is Vx. Observe that the horizontal vector Vx remains constant throughout the trajectory, but the vertical vector Vy is a shrinking vector in the upward course of travel and increases in length in the downward course of travel. The Vy vector determines the height of trajectory and the time of flight, where-as the Vx vector determines range. You are now ready to solve the problem.

Example (1a) Calculate the height of trajectory. First calculate the vertical component of velocity Vy.

Vy = (sin 0) (Vr)

Vy = (sin 30°) (100 f/s)

Vy = 50 f/s

Next use the acceleration formula to determine the time up.

A = (Vy)/(Tup)

Tup = (Vy)/g = (50 f/s)/(32 f/s^2)

(Remember: gravity is not only a force, but also an acceleration.)

Tup = 1.56 sec.

Use the free-fall equation to calculate the height.

H = (.5)(g)(Tup)----H = (.5)(32 f/s^2)(1.56s) = 38.94 ft.

(b) Calculate the total time of flight.

T = Tup + Tdown = 2(Tup)

T = 2(1.56 sec) = 3.12 sec

Calculate the range (R). First calculate the horizontal component of velocity (Vx).

Vx = (cos 0) (Vr)

Vx = (cos 30°) (100 f/s) = 86.6 f/s

The horizontal component of velocity is a constant and therefore the same as average velocity.

R = (Vx) (T) = (86.6 f/s) (3.12 sec)

R = 270 ft.

(d) The arrow would go over three times the distance to the target.

The example problem above illustrates the importance of lofting an object upward in order to gain distance. For example, all artillery guns are aimed upward in order to maximize distance. In throwing a pass, the quarterback knows that he must loft the ball at a substantial angle above the surface of the field in order to connect with a long pass. From experience, you know that there is an angle of elevation above which distance is lost and height is gained. Any person who has hit a high infield ball, or has hit under his golf ball, is very much aware of this fact. Bear in mind that all of these facts are relevant only as long as the object is lofted freely into motion, with only the force of gravity (a constant), with no other forces operating. Soon, you will begin the study of space trajectories in which different kinds of forces will operate to affect the paths of flight. Before you can begin the study of flight in space, you must learn about the laws developed by Sir Isaac Newton. His Universal Laws set down the conditions which govern motion and orderliness in the universe.

Exercise 10

Self Test

Solve as directed and show all important work.

100 mi/hr

(1) In throwing a baseball to person 100 feet away, you release the baseball eight feet above the ground, parallel to the ground, with a velocity of 100 mi/hr. (a) In how many seconds will it hit the ground if not caught? (b) What distance will it go prior to hitting the ground? (c) Was it caught before hitting the ground?

(2) A ball is lofted underhand, straight up in the air, with a velocity of 25 m/sec. (a) How high did it go? (b) What was it velocity at the top if its trajectory? (c) How long did it take to reach the top of its trajectory? (d) With what velocity did it pass the point of loft?

(3) A rocket is fired form its launch pad with a velocity of 2500 m/sec, at an angle of 70 degrees. (a) Calculate its horizontal and vertical components of launch. (b) Calculate the maximum height achieved by the rocket and the final range upon falling to the ground. (c) What was its final velocity and the direction of impact upon impacting the ground? (Assume the atmospheric conditions were frictionless.)

Newton's First Law of Motion

A body at rest or in uniform motion will remain so unless acted upon by an external force. This law affects every aspect of motion on earth. When you awaken in the morning, you must apply an external force in order to remove yourself from your position of rest. The glass of juice which is at rest on the table will not move to your mouth until you apply an external force to make it do so.

If you are coasting in a car or on your bike, you will continue in uniform motion unless you apply an external force by applying the brakes or by pressing on the gas peddle or pushing the peddles on your bike. One application will slow you down and the other will speed you up.

You are aware that it is much easier to push a person in a wagon than to physically push a car with a person in it. One is harder to push than the other because it is heavier than the other. This is true because the car has greater mass than the wagon. The inertia of a body is directly proportional to the quantity of matter which it contains. Mass and inertia are directly proportional and each is a constant in the universe. The metric units of mass are grams and kilograms. The English unit of mass is the slug. For the present time, accept the fact that 1 slug weighs 32 pounds; 1 gram weighs 980 dynes; 1 kilogram weighs 9.8 newtons.

Newtons's Second Law of Motion

When a net force is applied to a body, it is accelerated in the direction of the force with an acceleration that is directly proportional to that force, but which is inversely proportional to the mass of the body. For example: Place 50 pounds (p) in a wagon and push the wagon with a given force. The wagon will accelerate in the direction of the applied force with an acceleration (A). Next place a 100 p box on the wagon and push the wagon with the same force, the wagon will accelerate ½ as fast as before, ie., the acceleration will be (A/2). Now place a 25 p box in the wagon and push with the same force. The wagon will now accelerate at (2A) or twice as fast as before. Read Newton's Second Law again and see if it does not make more sense to you. (Assume the wagon to be weightless.)

The equations of Newton's second law are as follows: (1) Acceleration is directly proportional to the force.

Acceleration (a) $\propto$ Force (F) or a $\propto$F

(2) Acceleration is inversely proportional to the mass of a body.

Acceleration (a) $\propto$ 1/mass (m) or a $\propto$1/m

(3) These two statements combine into a single proportion as follows:

Acceleration $\propto$ Force/mass or a $\propto$F/m

These relationships can now be written as a universal equation in which:

Force = (mass) (acceleration) or F = (m)(a) or F = (m)(g)

The units of Newton's Second Law of Motion are listed in this chart.

Force	Mass	Acceleration	Force (composite)
Pounds	Slugs	Ft/sec^2	Slug-ft/sec^2
Newtons	Kilograms	M/sec^2	Kg-m/sec^2

Webster defines force as that which is capable of producing motion in a body, or which causes a change in its motion. Without the physical concept of force, you were severely limited in your ability to solve motion problems. Now you can apply forces and produce acceleration. Here are some example problems:

Example 1. (a) What constantly applied force will cause a 25 kg block of ice to accelerate 5 m/s^2?

F = (m)(a) = (25kg)(5m/s^2) = 125 kg=m/s^2 = 125N

(b) What constant force in pounds will cause a 1.72 slug block of ice to accelerate to 16.4 ft/sec^2?

F = (m)(a) = (1.72 sl)(16.4 ft/s^2) = 28.21 sl-ft/s^2 = 28.21 p

Example 2. (a) Calculate the weight of a 25 kg mass. (Remember that force and weight have the same units.)

F = W = (m)(g) = (25 kg)(9.8 m/s^2) = 24 5.N

(b) Calculate the weight of a 1.72 slug mass.

F = (m)(g) = (1.72 sl)(32 ft/s^2) = 55.04 p

Example 3. What is the acceleration of your 1420 p car if the applied engine force is 257 p? (Don't forget to convert the weight of the car to mass units.)

F = (W/g)(a)---a = (F)(g)/W=

a = (257p)(32 ft/s^2)/(1420 p) = 5.79 ft/s^2

Example 4. Assume your weight to be 140 p and that you are skiing down a thirty, 30° slope. (a) Make a scaled vector diagram which shows your weight and that part of your weight which acts parallel to the slope? (b) What force in pounds is acting to pull you down the slope? (c) Calculate the acceleration down the slope. (d) Calculate the component of gravity that acts parallel to the slope and compare this with your calculated acceleration from the diagram. (You may calculate all of these values- your answers should be very nearly the same.)

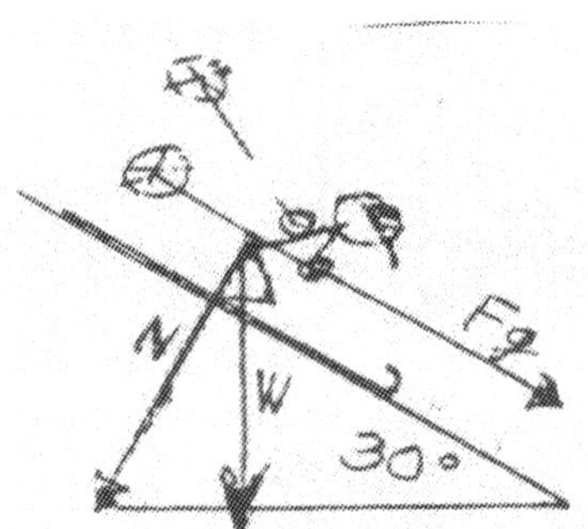

Exercise 11

Self Test

(1) A mass of 5000 grams is in free fall. What is the acceleration of this object?

(2) A weight of 145 p is in free fall. What is its acceleration?

(3) Calculate the number of slugs in 1 ton of weight.

(4) What is the weight of 67 slugs?

(5) Calculate the equivalent mass of 95 Newtons.

(6) A force of 15N moves an object weighing 47N. What is the acceleration of this object.

(7) A box weighing 500 p is pushed with a force of 72 p across a hard surface which has a frictional resistance to motion of 12 p. What is the acceleration of the box?

(8) A sled weighing 21 p is carrying a person who weighs 139 p, if their combined acceleration down a hill is 12 ft/sec-sec, what is the moving force if the surface is frictionless?

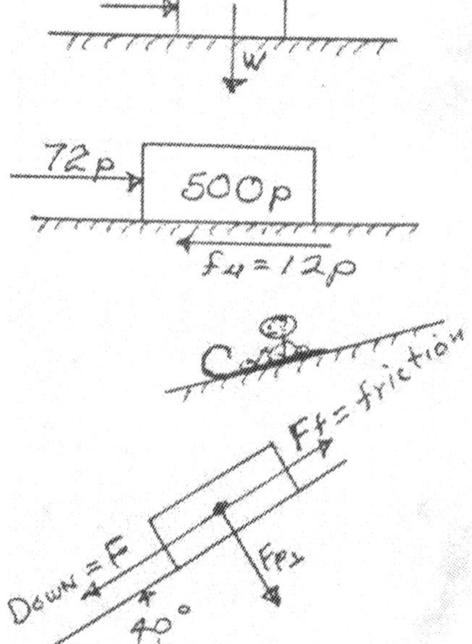

(9) A box weighing 300 p sets on a 40° ramp. (a) What force acts down the ramp? (b) What force acts perpendicular to the ramp? Make a scaled drawing of this situation similar to the sketch to the right. (c) If the resisting force of friction is 33 p will the box slide down the ramp? Which force is greater, the force down the ramp or the force of friction.

Motion in a Circle

Circular motion is an important part of everyday living. The wheels on your car form a circle and they move in a circle. It is circular motion at an amusement park that enhances the thrill of the ride, ie., the roller coaster and the Ferris wheel.

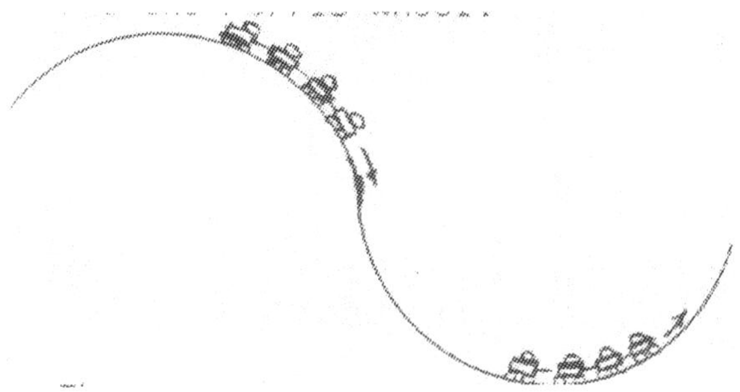

The same concepts are used in circular motion as in linear motion, but different symbols are used.

Distance	Velocity	Accel.
Linear θ D or S = θ Theta = θ	Linear As given V = ω Omega = ω	Linear As given a = ∝ Alpha = ∝
Circular Units		
Degrees Radians Revolutions	Deg/sec Rad/sec Rev/sec	Deg/sec^2 Rad/sec^2 Rev/sec^2
θ = ω x t	ω = θ/t	∝ = ω/t
1 radian = 180°/π = 57.32°		

When David slew Goliath, he used a sling shot which he whirled above his head. An object can be hurled with great velocity in this manner. The circle below can be used to represent the physics of this device.

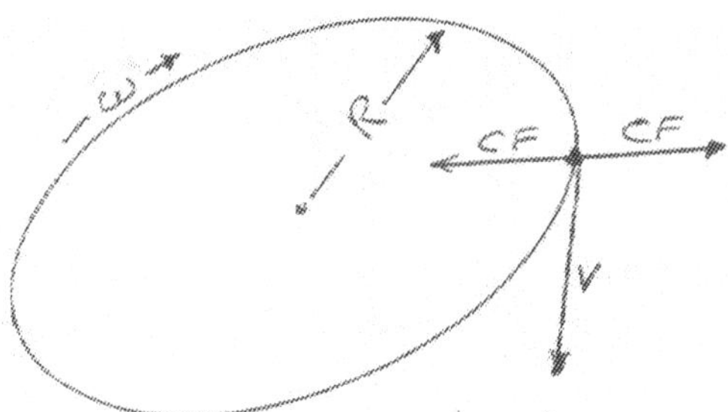

Assume that the thong used by David was four feet long and that he whirled it around and around until it was revolving at 60 revolutions in 20 seconds. In the circle drawn below, the radius is 4 ft and (ω) represents the angular velocity of the stone which he hurled. Then it follows that (v) represents the linear velocity of the stone. Ex. (1a) calculate the number of revolutions of the stone per sec.

ω = Rev/t = 60 rev/20s = 3 rev/sec

Frequency is defined as the number of rev/sec. Frequency = ω = 3 rev/s

The period (T) is defined as the length of time of one revolution.

The period T = 1/f = 1/(3 rev/s) = .33 sec/rev.

Calculate the linear distance the stone travels in one revolution. This is equal to the circumference of the circle.

D = C = 2πR = 6.28(4 ft) = 25.12 F.

Calculate the number of degrees (θ) traveled by the stone in 20 sec.

θ = (360°) x (60 rev) = 2.14 x 10^4 deg.

Calculate the linear velocity of stone as it moves about the circle:

V = (25.12 ft)(3rev/sec = 75.36 F/s

How fast in miles per hour would the stone be moving if it broke away from the string?

V = (.682)(75.36 f/s) = 51.4 mi/hr

Centri-Forces

There are two kinds of forces that act upon the rotating stone. Centriptital force is the inward force exerted upon the stone by the string. The counter force is the Centrifugal force which acts outward to balance the inward force. In uniform circular motion, the stone is moving with a constant velocity (v) which always acts tangent to the circle of motion. If the string breaks, the stone always moves away from the circle on a line tangent to the circle. Think of the stone as always being accelerated toward the center of the circle with an acceleration (a).

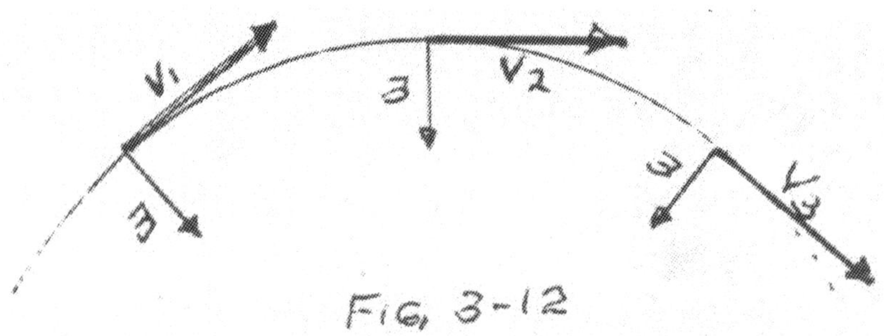

FIG. 3-12

Increase the parameters of this analogy to include earth and sun. What holds true for the string and stone, also holds true for earth and sun. But in this case, gravity is the string that holds the earth in orbit about the sun. You are very close to making calculations about velocities required to place rockets and satellites in orbits above earth. The interacting forces are the Centri-forces and the velocities and accelerations are those of orbiting vehicles.

These formulae can be derived by modifying Newton's second law.

F = ma, where (a) is acceleration, and (a) = $(\text{velocity})^2$/radius, then F = m(v^2/r)

Example (2) A lead wt. Of 0.1 slugs is swung horizontally overhead in a circle of radius 5 ft. If the velocity of the weight is 18 ft/sec, calculate the following: (a) The centripetal force.

F = (m)(v^2)/r = (.1s1)$(18\ f/s)^2$/(5ft) CF = 6.48 p

(b) Calculate the centripetal acceleration of the weight.

A = (v^2)/(r) = $(18ft/s)^2$/(5ft) = a = 64.8ft/sec^2.

(c) How many rev/sec does this velocity equal?

1 rev = C = 2πr = (6.28)(5 ft) = Rev/sec = (18 ft/sec)/(31.4 f/rev) Rev/sec = .573 rps.

Example (3) As part of their training, astronauts are swung in a great circle in specially designed equipment to condition them to the gravitational (g) forces of space travel.

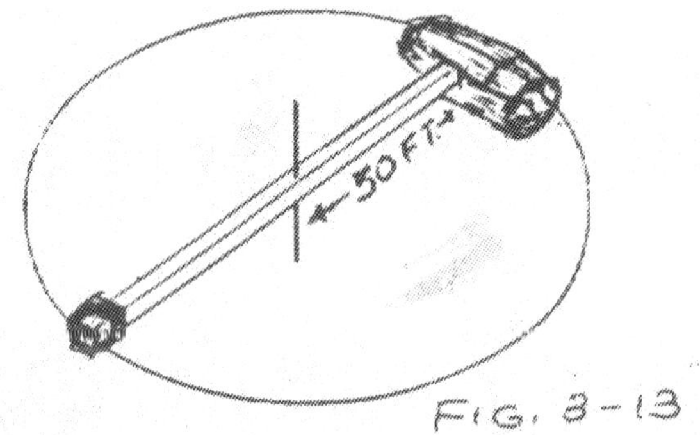

If the radius of the centrifugal force machine is 50 ft and he is swung at 12 rev/min, calculate (1) the circumference of the circle through which he swings; (2) his tangential velocity; (3) His radial acceleration (acceleration along the radius toward the center of the circle); (4) The number of (g) forces acting upon his body.

(1) $C = 2\pi r = 6.28(50\text{ft}) = 314$ ft.
(2) Rev/sec = (12 rpm)/(60 s/min) = rps = .2 rps
V = ©(w) = (314 ft)(.2 rev/sec) = 62.8 ft/sec
(3) $a = (v^2)/(r) = (62.8\ \text{f/s})^2/(50\text{ft}) = 78.88\ \text{ft/sec}^2$.
(4) (g factors) = $(78.88\ \text{f/s}^2)/32\ \text{f/s}^2) = 2.47$

This means there are 2.47 forces of gravity acting upon this person.

Exercise 12

Self Test

1. You are working on your bike which is upside down. You spin the 30 inch front wheel and count the number of revolutions of the wheel to be 90 revolutions for 30 seconds of time. Answer the following:

(a) What is the circumference of the wheel in feet?
(b) How many revolutions does the wheel make in one second?
(c) What is the linear velocity of a point on the rim of the wheel in (ft/sec)? This is the same as the linear velocity of the wheel.
(d) How many degrees does the wheel turn through in one second?
(e) How many radians does this equal?
(f) What is the frequency of revolution of the wheel?
(g) What is the period of turn, ie., how many seconds does it take to make one revolution?
(h) What is the tangential velocity of the wheel?
(i) What is the radial acceleration of the wheel?

2. You place your gloved hand on the rim of the wheel and slow it down to 1 rev/sec in 2 seconds.

(a) by how much has the angular velocity been reduced? (Answer in rev/sec and radians/sec.)
(b) What was the deceleration of the wheel in rev/sec-sec?

Moon Mathematics

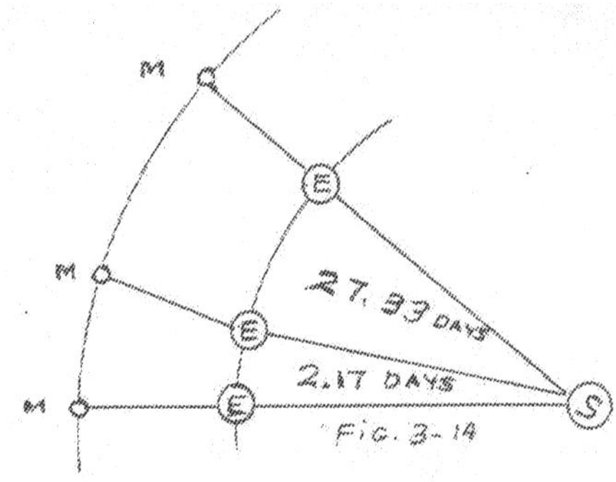

You have already learned more about the moon than most people know. Refer to page (8) for a review of this information. The moon is held in orbit by a very sturdy gravitational string. It takes the moon 27.33 days to complete its orbit or sidereal period about earth. However, because the earth is also moving around the sun at the same time, the time between each cycle of lunar phase is extended by two more days. This is called the synodic period which is 29.5 days. The above drawing will help you understand the difference between the two terms.

The following are some example problems that will help your understanding of the moon's periodic motion. Example (1) The mean distance of the moon from earth is 384,400 km. If the true period of the moon is 27.33 days, what is this period in hours, minutes and seconds?

Hours = (27.33) days)(24 hrs/day) = 655.92 hours
Minutes = (.92hr)(60 min/hr) = 55.2 minutes ----655:55:12
Seconds = (.2 sec)(60 sec/min) = 12 seconds
Example (2) Calculate the frequency of the moon about earth in rev/day and /hr.
Rev/day = 1/T = 1/(27.33 d/rev) = .0366 rev/day or Frequency
Rev/hr = 1/T = 1/(655.92 hr/rev) = 1.52×10^{-3} rev/hr
Example (3) Calculate the orbital distance the moon travels around earth in kilometers and miles.
D = Cir. = $2\pi R$ = (6.28)(384,400km) = 2.414×10^6 km
D = (2.414 x 10 km)/(1.6 km/mi) = 1.51×10^6 miles
Example (4) What additional linear distance in kilometers and angular distance in degrees does the moon travel in 2.17 days in order to make up its synodic period?
D = (29.5/27.33) x (2.414 x (2.414 x 10^6 km) = 2.606×10^6 km
$\Delta D = (2.606 - 2.414) \times 10^6$ km = 1.92×10^5 km
$\Delta\theta = (29.5/27.33) \times (360°) - 360° = 28.58°$

1. Mars has two moons. Deimos, the farthest, is 23,500 km from Mars and makes one revolution about the planet in 1.26 days. (a) How many hours, minutes and seconds does this equal? (b) What frequency in revolutions per hour does this equal? (c) Calculate the orbital distance traveled by Deimos in making one revolution around the mother planet. (d) Calculate the velocity of Deimos in Mi/hr as it moves through its orbit.

Newton's Third Law of motion

Newton's Third Law states that for every action there is an equal and opposite reaction force. For example, if you weigh 150 p and you are sitting on a chair, you push upon the chair with a force of 150 p and it pushes back upon you with the same force. Another example is that of pushing upon a heavy door to open it and having it push back upon your hands with a nearly equal force. In the case of the chair, you would fall to the floor if it were not strong enough to support your weight. A heavy door may cause you to move backward when you push upon it. The final example has to do with rocket propulsion. Rockets are propelled forward due to the velocity of gas particles being ejected from the engine. The action is the flow of particles out of the engine. The reaction is the motion of the rocket moving forward. The greater the velocity of ejection, the faster the rocket moves through space.

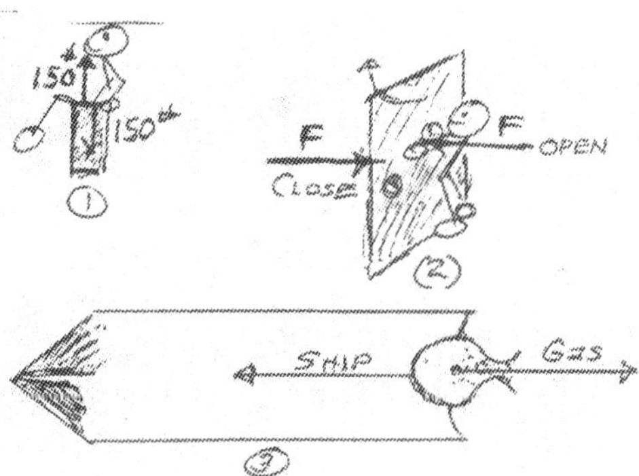

Newton's Universal Law of Gravitation

Newton's Gravitational Law states that every object in the universe attracts every other object with a force directly proportional to the square of the distance between their centers. In simple terms, this means that every object in your home, your place of work, including you is attracted to every other object through the medium of gravitational forces. The vehicle of these forces is the "graviton." Its energy, though hypothesized, has not been captured or measured. These forces are very weak. Only very large masses produce gravitational fields that can have an observable effect upon us. Gravitational forces have been measured in the laboratory.

Newton's gravitational law is written in formula form below and illustrated in the following drawing:

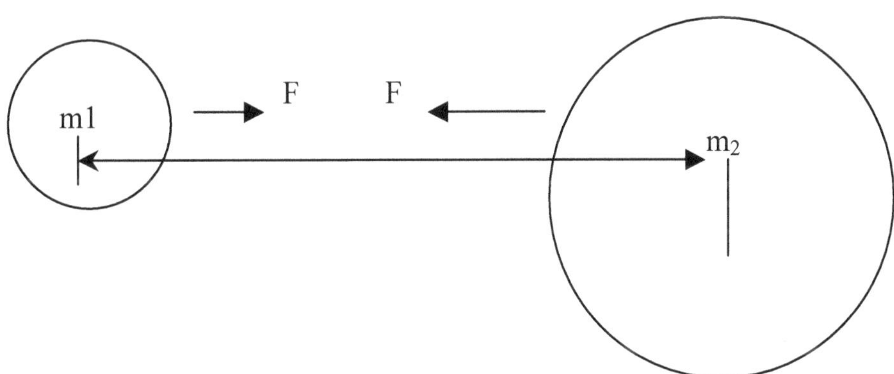

$$F \alpha \frac{(M_1)(M_2)}{R^2} = \frac{G(M_1)(M_2)}{R^2}$$ Where G = the universal gravity constant

F=Newtons G= $(6.66 \times 10^{-11}) \frac{M}{kg \sec^2}$

Fig 3-16

In order to change the proportion into a valid operational equation, it is necessary to place a constant in the equation. The development of this constant was Sir Isaac Newton's great contribution t world science. The most remarkable attribute of this constant is that it holds true throughout the universe, ie., in all measurements made by Astronomers for all times.

This is the metric value of G. Could you derive the English value if you were given time to do so? The solution is sequentially done as follows:

Change cubic meters to cubic feet. $(m)^3 \times (3.281\ ft/m)^3 = 35.32\ ft^3$
Next change kilograms to slugs. 1kg = 2.2 p ------- 2.2p/kg
Covert 2.2p into slugs. $2.2p/32ft/sec^2 = .0688$ slugs

$$(G)\ English = (6.66 \times 10\text{-}11)\frac{(35.32 ft^3)}{(0.688 sl - \sec^2} = 3.41 \times 10^{-8}\ \frac{ft^3}{sl - \sec^2}$$

Example 1. In a factory, two large steel ingots weighing 5 tons each, are setting side by side with their centers ten feet apart. Calculate the force of gravitational attraction between the two ingots. The solution is a s follows:

First calculate the mass of one ingot in slugs.
Mass = weight/gravity = $(5T \times 2000\ p/T)/\ (g) = 312.5\ p/ft/s^2 = 312.5$ slugs

$$F = \frac{G(312.5 sl)^2}{(10 ft)^2} = \frac{G(97545.3 slugs^2)}{100 ft^2} = 976.56\ sl^2/ft^2$$

$F = (3.41 \times 10^{-8} ft^3/sl\text{-}sec^2)(976.56\ sl^2/ft^2)$

$F = 3330 \times 10^{-8}$ sl-ft/sec $= 3.33 \times 10^{-6}$ pounds

Determination of Gravity in Space

For purposes of orbiting vehicles, ie., satellites and rockets, it is important to be able to calculate gravity above the surface of earth. Newton's gravitational law can be used to develop a simple equation that will permit this determination. Given F = (m)(g), where (m) is an object on the surface of earth, this can be equated on the left side of Newton's gravity equation as follows:

$$F = (m)(g) = \frac{G(m)(Me)}{R^2}$$

Divide both sides of the equation by (m).
The equation becomes

$$G = \frac{G(Me)}{R^2}$$, where (Me) is earth's mass.

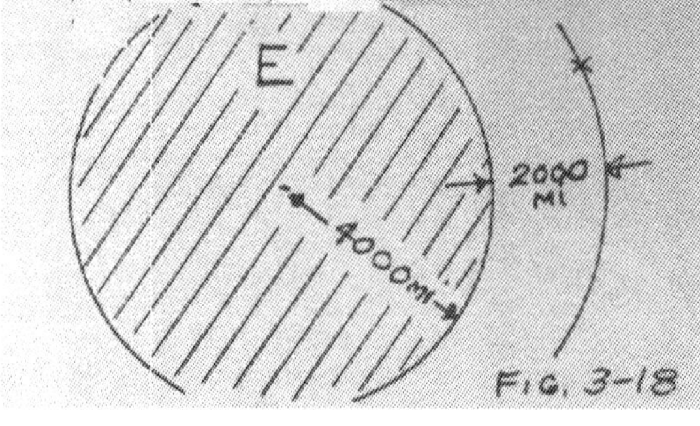

FIG. 3-18

The equation demonstrates that gravity is inversely proportional to the square of the distance from the center of earth. Not only can the equation be used to calculate gravity at any point above the surface of earth, but it shows why all objects on earth fall with equal acceleration. (What about the equation shows this important fact?) Example (20 Calculate gravity at a point 2000 miles above the surface of earth. (6000 mi = 9600 km)

$$g = G(Me)/R^2 = \frac{G(5.97x10^{24}\,kg}{(9.6x10^6\,m)^2} = (6.66x10^{-11}\,kg - m^3/s^2)(6.48x10^{10}\,kg/m)$$

g = 4.31 m/sec-sec

Excrcisc 14

Self Test

1. Two steel battle ships are docked side by side in a ship yard. If each weights 7.3 billion pounds and their centers are 750 feet apart, what force is acting to pull them together? Do you think they would ever drift close enough together to make contact with each other if left free to do so?
2. A satellite is placed in orbit 750 miles above the surface of the earth. Calculate the acceleration of gravity at this point.

Chapter IV

Production of Motion

Energy and Work

Imagine a world existing in a state without the benefit of energy. Such a body would have no life and no dynamic. It would exist as a hunk of ice far from any star or sun such as our own. Energy is the source of life in our universe and when utilized for the benefit of mankind, under his control, provides for his pleasure, comfort and industry. Energy is that form of matter which makes work possible, or, energy is simply defined as the ability to do work. It is man's ability through work, to move other matter and to build, simply defined, work is done when a force moves through a distance, ie., work is (force) x (distance). If you attempt to push a car by exerting a force upon it, but the car does not move, there has been no work done. The units of work in the (mks) and (cgs) systems and the (fps) are expressed as follows:

WORK = FORCE X DISTANCE

Work (mks) = Newtons x meters = Nt-meters = Joules

Work (fps) = Pounds x feet = Ft-pounds = F-p

Work (cgs) = Dynes x centimeters = Dyne-cm = Ergs
(Cgs units are for your information & too small for your work)

Example (1) If you exert a force of 35 p to move an object 20 feet, how much work have you done?
W = Fs or (35p) x (20 ft) = 700 f-p of work

Example (2) You lift a box weighing 25 N from the floor placing it on a table 0.8m high. How much work did you do?
W = Fs or (25N) x (.8m) = 20 Joules of work.

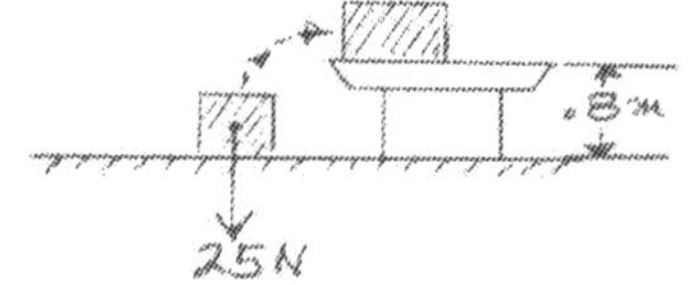

Frictions

Millions of Joules or Ft-p of work are done when a rocket moves from its pad into the atmosphere and out into space. Fantastic quantities of energy are expended in heat alone. Most meteors that fall from outer space burn up before they reach earth. It is the friction of the atmosphere that heats them to incandescence, that causes them to emit light and heat. It is this friction that will cause your re-entry vehicle to heat to over 5000 degrees when your return from the moon. The greater the density of the atmosphere, the more your vehicle will slow down due to molecular friction. This force of friction is directly proportional to the square of your velocity as you enter the earth's atmosphere. This is a good aspect of friction when it serves your purpose. For the most part, friction is a negative aspect of work. It slows you down when you want to go faster, it causes your car to use more fuel than you would like, and it wears out wheel bearings that must be replaced.

Mechanical friction results when two hard surfaces, usually one stationary and the other moving, or about to move, interact against each other. This type of friction is either starting friction or moving friction. Rolling friction is similar to sliding friction, but is quantitatively less. Fluid friction is another kind of friction, which results when liquids and gases interact with solids passing through them or with solids they pass through. <u>In the case of mechanical friction, the force to overcome friction is directly proportional to the normal force acting between the two bodies.</u> It also depends upon the composition and smoothness of the two surfaces. The formula relationship is expressed as follows:

(F) friction force = (u)(N) normal force or Fu =(u)(N)

(u) is the constant of friction which has to do with different types of surfaces, ie., metal to metal, wood to wood, plastic to plastic or an inter-mixing of such materials. In the case of gases and liquids, it is the density factor that makes the difference in the operation of frictional forces.

Example (1) IF the coefficient of friction between a 50 p box resting on the floor is (.21), calculate the force needed to overcome friction.

Frictional force = (u)(N) = (.21)(50 p) = 10.5 pounds to overcome friction

Example (2) Reference Ex. (1). If a 25 p force pushes the box across the floor for 12 ft, how much work was done against friction? What total amount of work was done?

Work(u) = Fu x D = 10.5 p x 12 ft = 126 ft-p against friction

Work(tot) = F x D = 25p x 12 ft = 300 ft-p

Example (3) If a Titan rocket weighs 60 tons, how much work is done by its engines in lifting it the first ten feet off of the pad? (Do not consider the amount of fuel expended when you make your calculation.)

W = (60 tons)(2000 p/ton)(10 ft) = 1,200,000 ft-p

Example (4) Reference Ex. (2): If the force was 25 N and the distance was 1.2 meters, how much work did you do in Joules?

W = (25 N) x (1.2 m) = 30 Joules of work

Exercise 15

Self Test

1. If 530 N-m of work are done in lifting an object, how many Joules of work are done?
2. A force of 40 p is required to set a 300p cart into motion.
   (a) What is the coefficient of starting friction.
   (b) If a force of 30 p is required to maintain a constant speed, what is the coefficient of moving friction?
   (c) If the cart moves through a distance of 275 feet, how much work is done against friction?

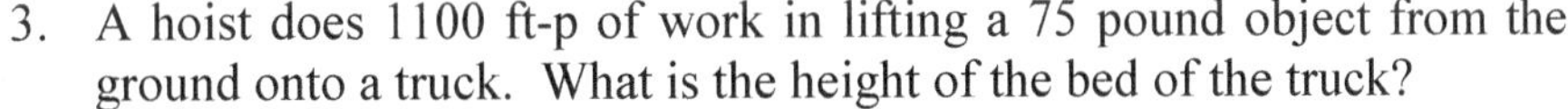

3. A hoist does 1100 ft-p of work in lifting a 75 pound object from the ground onto a truck. What is the height of the bed of the truck?
4. An orbiter enters the lower stratosphere with a velocity of 12,000 mi/hr. If a given density of atmosphere reduces its velocity 100 mi/hr/sec, (a) How long will it be until its velocity is 6000 mi/hr? (b) If the density of atmosphere increases by a factor of 3x, after 6000 mi/hr, how long until its velocity is zero?

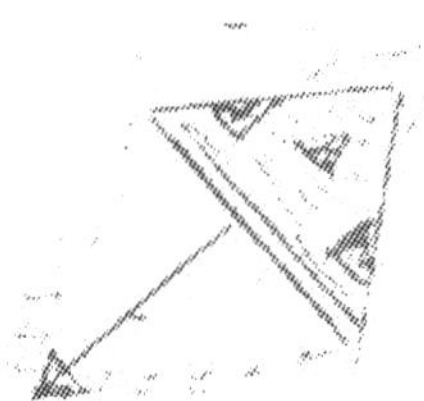

5. A 500 p box rests on a surface inclined at 40°. (a) How much weight acts normal to the plane? (b) How much weight acts parallel to the plane? (c) If the coefficient of friction between the surfaces is 0.35, what is the force of sliding friction? (d) What applied force will keep it from moving down the plane?

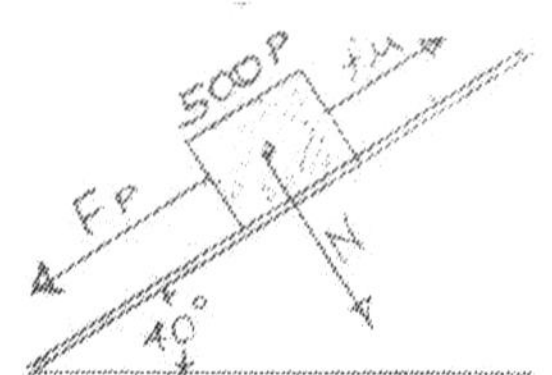

The Law of Conservation of Energy

The Law of Conservation of Energy states that energy can neither be created or destroyed. At any one time, the total of all mass and energy in the universe is a constant. With the advent of the nuclear age, ie., the first Fermi produced nuclear fission of matter and the resulting energy produced, this law became the extant in time Einstein's prediction that matter could be converted into energy in keeping with his equation $E=mc^2$, achieved universal acceptance. Already nuclear engines are the driving force for submarines and large ships at sea. Modifications of these engines will soon be the power units used in space travel to distant planets and places in this galaxy.

Potential Mechanical Energy

Your book is on your desk and your arm brushes it knocking it to the floor. While on the desk, it had potential energy with respect to the floor. This is called energy of position. You know it had energy relative to the floor, because it made a loud noise when it hit the floor. Had your arm only moved it a few inches on the surface of the desk, there would have been little or no perceived noise. The quantitative energy of the book on the floor is less than at the top of the desk, because you must do work on the book to move it back to the top of the desk where its energy will be the same as before. Potential Energy is energy due to position or to configuration. If you wind a clock, the spring has potential energy due to configuration and will operate the clock for many hours.

The formula for potential energy is as follows:

(PE) potential energy = mass x gravity x height.

PE – (m)(g)(h) = work

Example (1) Assume your book has a mass of .0625 slugs and that your desk is 2.5 feet high. Calculate the potential energy of the book while on the desk.

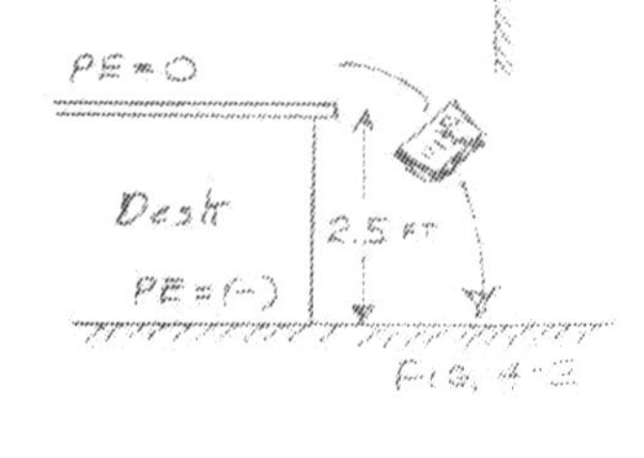

PE = mgh = (.0625 sl)(32 ft/s^2)(2.5ft) = 5 ft-p of energy

This is exactly equal to the amount of work you must do to retrieve the book and place it back on top of the desk.

Work = (W) force x D = (2p)(2.5 ft) = 5 ft-p of work

Take note of the fact that the units of energy are the same as those for work in all systems. Also notice that energy was conserved in this mechanical situation. If the floor is considered to be zero potential for the book, then the PE of the book on the desk is positive. If the zero reference for PE as the desk, then the books PE on the floor would be negative (-5 ft-p). However, if the book were then placed on a shelf 2.5 feet above the desk, that would be (+) potential energy for the book.

Mechanical (Kinetic Energy)

From a mechanical perspective, it can be said that potential energy is conserved when all of its energy is converted into kinetic energy. If PE is the energy of position, then KE is the energy of motion. By definition, kinetic energy is the ability of a body to do work due to its quantity of motion. For this reason, when your book fell off of the desk, the energy that it possessed on the desk, was converted to kinetic energy as it fell from the desk. Since this energy is conserved in a theoretical sense, if energy losses due to heat are not considered, then the law of conservation of energy also holds true for this transformation of mechanical energy.

Potential Energy = Kinetic Energy = Work Done
Kinetic Energy = ½ (mass)(velocity)^2 = ½ (m)(v^2)
Energy = (m)(g)(h) = ½ (m)(v^2) = (F)(D)

As with potential energy, the units for kinetic energy are the same as for work. Then energy equations permit you to easily solve problems in physics that otherwise would be very difficult.

Example (1) A one ton car is traveling at 60 mi/hr, calculate its KE.

KE = ½ (m)(v^2) = .5(2000 p/32 ft/sec^2)[(60 mi/hr)(5,2380 ft/mi/3600s/hr)]^2

KE = 242,000 ft-p

Example (2) If the car in number (1) coasts to the top of a hill and comes to rest there, what will be its potential energy?

Example (3) A dam is 0.1 km high. If a metric ton of water flows down through a pipe to a turbine located at ½ the height of the dam, how much energy is available to the turbine to produce electricity?

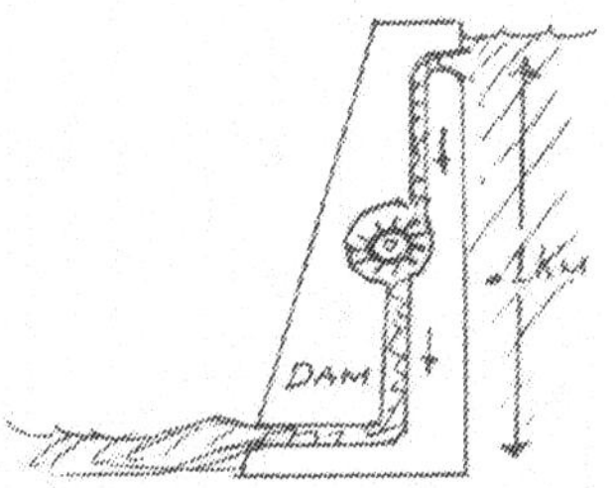

(b) With what velocity does the water reach the turbine?

(a) PE = (m)(g)(h) = (1000 kg)(9.8 m/s^2)(50m) = 490,000 Joules

(b) KE = PE = ½ (1000 kg)(v^2) = 490,000 Joules

$$v = \sqrt{980m^2/s^2} = 31.3m/s$$

(c) Assuming that all of the KE was spent in turning the turbine, how much energy remains to do work at the bottom of the dam? 490,000 Joules

A useful formula can be derived from the KE, PE relationship that can be used to solve problems similar to (3c) above.

KE = PE --- ½ (m)(v2) = (m)(g)(h) ----v2 = 2(g)(h)

$$v = \sqrt{2(g)(h)}$$

Mechanical Power

Power is the rate at which work is done. The faster you do a given job, the greater the application of power. When people speak about paying for electric power, they really mean electric energy. In a sense, it is wrong to speak of the "electric power company," but more correct to say the "electric energy company". Power is work done per unit of time. For this reason, the units of power differ from those of work and energy. The units of power are as follows:

Metric Power Units = Work/time = Joules/second = Watts
English Power Units = Horse-power = 550 ft-p/second
(1) horse power = 746 watts

Example (1) Three different persons do 350 Joules of work. (a) What is the power output of each person, if, respectively, they do the work in (1 sec), (2 sec) and (3 sec)?

(1st) person = 350 J/(1 sec) = 350 Watts of power
(2nd) person = 350 J/(2 sec) = 175 Watts of power

(3rd) person = 350 J/(3 sec) = (You supply the answer)
(b) Which of the three persons is the most powerful?
Example (2) A hoist can lift 1 ton a distance of ten feet in 5 sec.
(a) Calculate the power output in (ft-p/sec).
(b) How many horse power does this equal?
(c) How many Watts?
(d) Kilowatts?
(a) Power = work/time = (2000 p)(10 ft)/(5 sec) = 4000 ft-p/sec
(b) Horse Power = work/(550 ft-p/sec-HP = 7.27 HP
(d) Kilowatts = Watts/(1000w/kw) = 5.423 kw

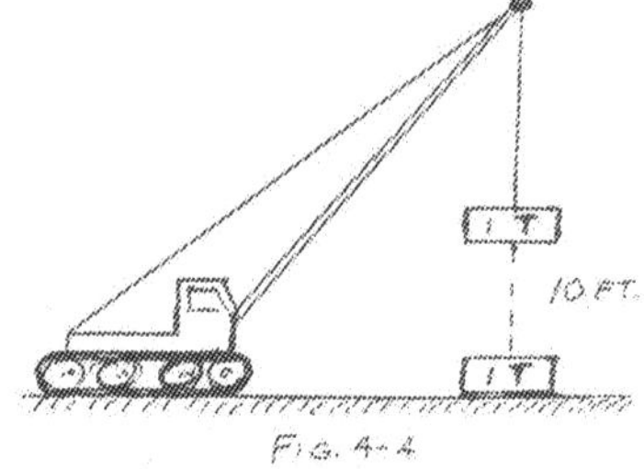

FIG. 4-4

Exercise 16

Self Test

1. A man and balloon weighing one half ton hover 500 feet above the ground. Calculate their PE.
2. A man of mass 7.3 slugs, stands at the top of stairs. Calculate his PE if he is 20 feet above the floor below.
3. Use energy equations to derive a formula which relates velocity and height.
4. Calculate the KE of a 12p bowling ball thrown with a velocity of 28 mi/hr. (Hint: you must convert the velocity to (ft/sec).)
5. If the man in the gondola in Problem (1) jumps, with what velocity will he hit the ground?
6. What is the KE of a 1500kg automobile traveling with a velocity of 72 km/hr?
7. If the engine in your car produces a constant force of 575 p to move the car 400 ft in 30 seconds: (a) How much work is done in moving the car that distance and (b) what horse power did the engine develop?

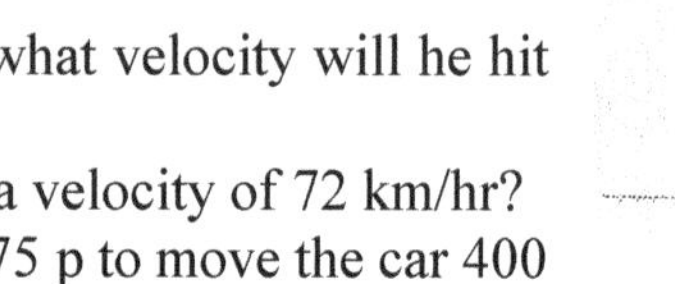

Momentum

Momentum may best be thought of as the quantity of motion which is had by a moving object. The subjective side of this (your feelings) tell you that when a big semi truck passes you on the highway, it has a great quantity of motion due to its velocity and size. You have the same feelings about a rumbling train that passes by while you wait at a RR Crossing. This is particularly true when you are in your own car which is small by comparison. Momentum is defined as the product of mass and velocity. Example (1a) Calculate the momentum of a 20 ton truck traveling at 6 ft/sec.

$$M = m \times v = \frac{(20T)(2000p/T)(66ft/\sec)}{32ft/s-s} = 8.25 \times 10^4 \text{ sl-ft/sec}$$

(b) Calculate the momentum of your 1500 p car traveling at the same velocity.

$$M = \frac{(1500p)(66ft/\sec}{32ft/s-s} = 3.094 \times 10^3 \text{ sl-ft/sec}$$

(c) How much greater is the momentum of the truck than that of your car? You can see why trucks pose so much danger to the passengers in an automobile when there is an accident.

The study of collision phenomena assumes that all collisions are perfectly elastic. For example, highway collisions are inelastic because of the bending of metal and resulting heat. The Law of Conservation of Momentum states that the total momentum before a collision event is equal to the total momentum after the event.

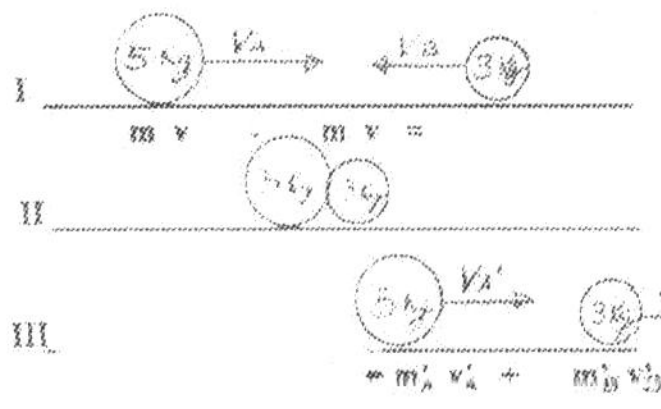

(I) Before: The two bearings A and B to the right are moving toward Each other on a collision course.

$M_A v_A - m_B v_B =$

(II) At collision: They then collide with each other.

(III) After collision, they move away from each other in their final respective directions.

$= m'_A v'_A + m'_B v'_B$

Problem (2) If their respective masses are 5 kg and 3 kg and their respective velocities before impact are 7 cm/s and 5 cm/s, and if after collision the (B) ball is moving away to the right with a velocity of 3 cm/s, calculate the velocity of the (A) ball.

$m_A v_A$	$m_B v_B$	$m_A v_A$	$m'_B v'_B$
(5kg)(7cm/s)	-(3kg)(5cm/s) =	(5kg)(v) +	(3kg)(3cm/s)
35kg-cm/s	-15kg-cm/s =	(5kg)(v) +	(9kg-cm/s
	$(5kg)v_1 =$	20kg-cm/s -	9kg-cm/s
	$V_A =$	2.2cm/s	

Was momentum conserved? Yes!! The momentum before the collision was [(35kg-cm/s)-(15kg-cm/sec)]. Afterwards the momentum was [(9kg-cm/s) + (11kg-cm/s)].

Impulse

The algebraic sum of the momenta of the two balls before and after collision was a constant even though the momenta of each changed after collision. What force operated to produce this change and over what period of time did it take place? The answers to these questions involve a new concept called 'impulse.' Impulse is the product of force and time, (F x t), and is equal to the change in momentum of either of the objects.

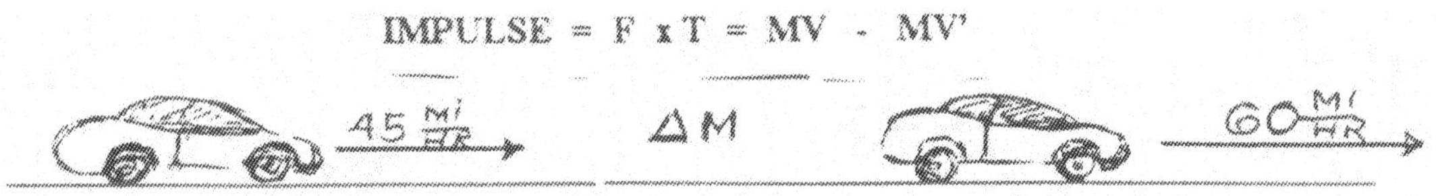

IMPULSE = F x T = MV - MV'

Example (3) You are driving your 1400 p car at a constant velocity of 45 mph. You step down on the gas peddle and accelerate to 60 mph in 5 seconds. Calculate (a) Your change in momentum; (b) The impulse produced by the engine; (c) The force that produced this change in momentum.

(a) Change in M = $\Delta M = m(v - v) = \dfrac{(1400p)((66 ft/s) - 44 ft/s))}{32 ft/s - s} = 962.5$ sl-ft/s

(b) Impulse = ΔM = 962.5 sl/ft/s (c) Force = F = $\Delta M/t$ = (962.5 sl-ft/s)/5 sec = 192.5 p

This aspect of the study of momentum is important because the concept of impulse provides you with a mathematical means enabling you to analyze the forces produced by burning fuels and the resulting velocities imparted to the rocket vehicle.

Example (4) You are in a space orbiter vehicle orbiting earth at a velocity of 10,000 mph. You fire your retro rocket to slow down to 6000 mph in order to enter a lower orbital pattern. If the engine fires for one minute, and your vehicle weighs 2 tons, calculate the following: (a) Your change in momentum; (b) The resulting impulse; (c) Your deceleration; (d) The force applied to the vehicle by the engine to slow it down.

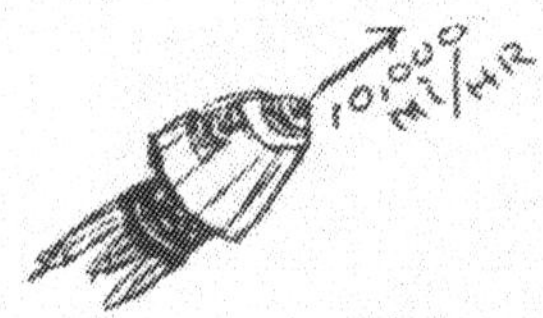

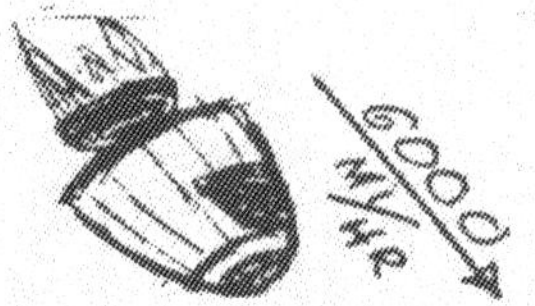

ΔV = 4000 mi/hr

(a) $M = mv\text{-}mv' = \dfrac{(4000p)(4000mph)(6{,}280 ft/mi)}{(32 ft/s-s)(3600 s/hr)} = 733{,}333\ sl\text{-}ft/s$

(b) Same as (a).

(c) $a = v/t = \dfrac{(-4000mph)(1.47)}{60\sec} = -.98$ ft/s-s

(d) $F = M/t = \dfrac{733{,}333\ sl\text{-}ft/s}{6} = 12{,}222.17$ p of force

Exercise 17

Self Test

(1) Two RR cars each weighing 2.5 tons collide and connect upon impact. If one of the RR cars was moving with a velocity of 15 mi/hr and the other one was at rest, calculate the following: (a) The momentum of each before impact. (b) Their combined velocity after impact. (c) Their combined momentum after impact. (d) Was momentum conserved? (Momentum answers are to be in slug-ft/sec.)

(2) A one ton howitzer fires a fifty pound shell. If the muzzle velocity of the shell was 1800 ft/sec, what was the velocity of recoil of the gun?

(3) A test rocket engine is mounted on a car on a RR track. If the engine and car weigh five metric tons and the engine accelerates both from rest to a velocity of 100 km/hour in 15 seconds, calculate the following: (a) The acceleration of the car and engine. (b) The final momentum of both in (kg-m/sec). (c) The impulse imparted to both by the engine. (d) The force applied through the engine.

(4) A one ton orbital vehicle orbits 75 miles above earth with a velocity of 8000 mi/hr. (a) What is its PE relative to earth? (b) What is its KE? (c) What is its total energy?

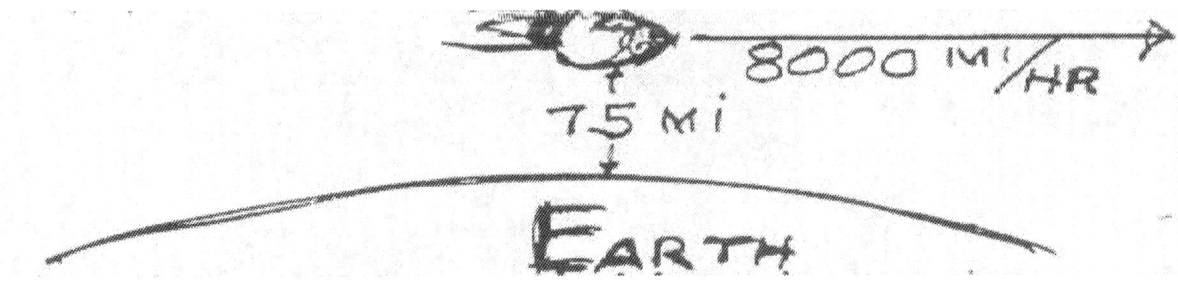

(5) The orbiter in the previous problem is moving parallel to the surface of the earth. If it reverses direction so that its thruster engine can slow it down and it fires for sixty seconds with a force of 2000 p, calculate the following: (a) The impulse reaction on the orbiter . (b) The final velocity of the orbiter. (c) If it fell out of the sky at this point, with what velocity would it impact the ground?

Chapter V

Satellite Orbits

Modes of Orbit

The path of the earth or any other planet about the sun is an orbit. Also, an orbit is the path of a man-made satellite about earth or any other heavenly body. Orbits can be perfectly circular or they can have elliptical "eccentricity". The eccentricity of closed orbits is between 0 and 1. A circle has an eccentricity of 0, whereas a parabolic escape orbit is given the eccentricity 1. A hyperbolic escape orbit is greater than 1. (ref. A)

What is an Ellipse?

Other than the fact that the planets move elliptically about the sun, what is an ellipse? An ellipse is the path of a point which moves so that the sum of its distances form two fixed points is constant. Ellipses are easily drawn. Place two pins about 4 inches apart on a thick piece of cardboard. Tie a knot in a piece of string about twelve inches long. Place the string around the two pins and trace out your ellipse as shown in (B).

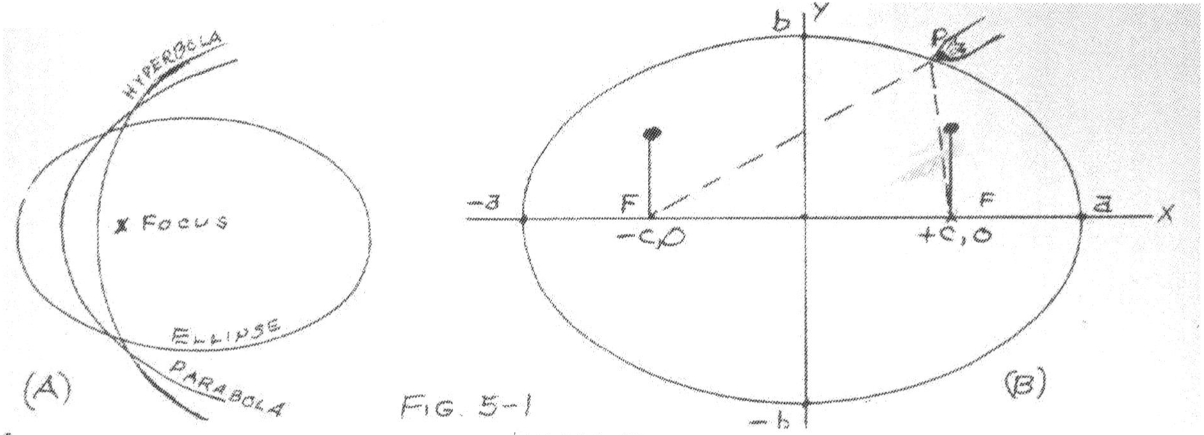

FIG. 5-1

If most of the objects in the universe move in elliptical orbits, even sub-atomic particles, ie., the electron around the nucleus of the atom, then you should learn some information about the ellipse.

The general equation of an ellipse (the same as (b) above) is as follows:

$$\frac{x^2}{a^2}+\frac{y^2}{b^2}=1$$

Even thought this equation may appear to be very complex, actually it is amazingly simple and much can be learned about the ellipse by a simple investigation of the letters. First notice that all of the letters are labeled in the drawing. Twice (a) is the length of the major axis and twice (b) is the length of the minor axis in the drawing. The coordinates of the foci (focal points) are found by taking the square root of the difference between the squares of (a) and (b).

Example (1) Given the following equation of an ellipse, make the following calculations.

$\frac{x^2}{25}+\frac{y^2}{16}=1$ change tot his form $\frac{x^2}{(5)^2}+\frac{y^2}{(4)^2}=1$

(a) Find the length of the major axis if the units are in centimeters. 2 x 5cm = 10cm.
(b) Find the length of the minor axis. 2 x 4cm = 8cm.
(c) Find the location of the focal points. $\sqrt{25cm^2-16cm^2=9cm^2}$. The focal points are at +3cm and –3cm.
(d) Use this information to sketch the ellipse. Simply locate the (+) and (-) points for the major and minor axes, and draw the ellipse.

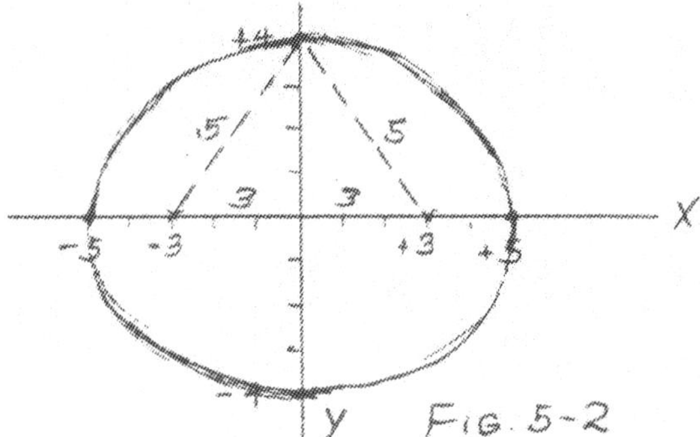

FIG. 5-2

(e) How long is the string? As shown in the drawing, the string must pass around the focal points. Complete the isosceles triangle as shown in the drawing and add up the lengths of the sides. Length = 5 cm + 5 cm + 6 cm. Length = 16cm.

Now it is easy to see that orbits can be very long and narrow, ie., highly eccentric, or they can be short and fat—just a little bit eccentric. You can create the type of ellipse you want to design by changing (a) and (b) in the equation.

The Application of Kepler's Laws to Elliptic Orbits

The mathematics of the ellipse was well known in Kepler's time. Kepler's genius was that he studied the orbits of the planets very carefully and could clearly see their orbits were not circular. Upon applying his conclusions in the form of his laws, his ideas gained universal acceptance.

Law I (The Law of Elliptic Orbits) Each planet moves in an elliptic orbit around the sun. The sun occupies one of its two foci as shown in (3).

Law II (The Law of Equal Areas) The imaginary line connecting any planet to the sun sweeps out equal areas of the ellipse in equal intervals of time. (a) When farthest from the focus (S), the planets move the slowest. (b) When nearest to the focus, they move the fastest.

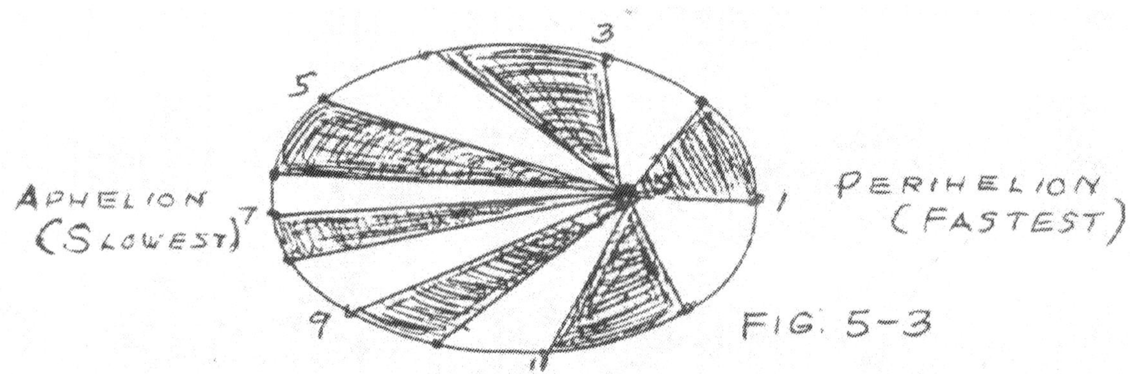

FIG. 5-3

Law III (Law of Periods) The square of any planet's period of orbital revolution is proportional to the cube of its mean distance from the sun.

To apply Kepler's Laws to orbiting satellites about earth, substitute "satellite" for planet, and "earth" for the sun.

Example (2) Given ($P^2 \propto D^3$), which is the formula statement of Law III above, calculate Jupiter's period about the sun. Jupiter's distance from the sun is 5.2 times that of earth or 5.2 AU.

$P^2 = (5.2)^3$ --- $P = \sqrt{140.61}$ = 11.86 times the period of earth which is one year.

Other Types of Orbits

There are many types of orbits. The inclination of orbit is the angle the orbit makes with the equator. An orbit that parallels the equator is one at zero degrees. A polar orbit is angled at 90 degrees. The higher an orbit the greater the length of time to pass around earth. The geosynchronous orbit about earth is a special orbit in which the object is set 22,300 miles above earth. At this distance from the center of earth, the period of orbit is 24 hours. If these orbits are made stationary with respect to earth, they are called geostationary orbits. This is the same as the period of earth's orbit. This is particularly good for communications satellites because earth bound antenna can be pointed at such an antenna and not because earth bound antenna can be pointed at such an antenna and not have to be moved. Sun-synchronous orbited satellites pass over every place on earth at the same time of day. These low orbits are set at about 98 degrees.

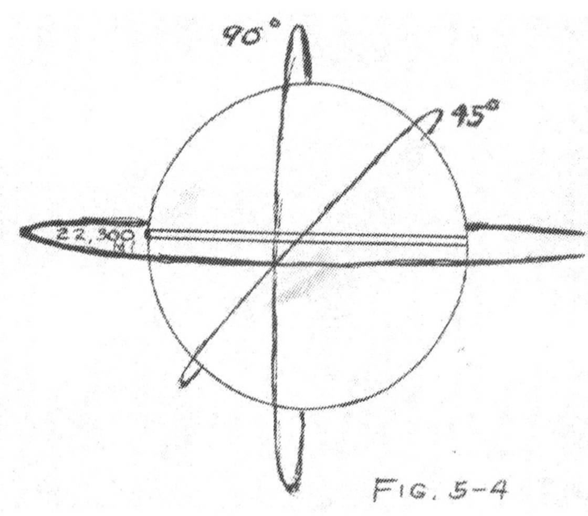

FIG. 5-4

Example (3) Calculate the orbital velocity of a satellite positioned for fixed orbit. What do you think it should be?

$$\text{Given } V = \sqrt{(g)(\text{Re})^2/(Rx)} = \sqrt{\frac{(32 ft/\sec)(4000 mi x 5280 ft/mi)^2}{26300 mi x 5,280 ft/mi}} =$$

$$V = \sqrt{\frac{1.43x10^{16}}{1.39x10^{8}}} = \underline{10,135 \text{ ft/sec or } 6,910 \text{ mi/hr}}$$

C (at 26300 mi) = (6.28)(26300mi) = 165,164 mi (Divide this number by V = 6910 mi/hr and you get 23.9 hours for your answer.) This is close enough to 24 hour (any error is due to rounding).

Orbiting Satellites

Now you are ready to do the mathematics needed to place men in orbit high above earth. Each time NASA orbits the space shuttle for another series of space experiments, you will be able to make the same calculations regarding the craft's space data that are made by NASA's engineers. Included in these calculations that pre-determine orbital travel are the equations having to do with gravity at orbit and the required velocity to hold the craft in orbit. Once again, you must deal with Newton's "force" and "gravity" equations in order to derive these formulae. These equation work as follows: Divide each pair of equations.

$$F = (m)g_E \propto \frac{(m)(Me)}{(\mathrm{Re})^2} \rightarrow g_E = \frac{G(Me)}{(\mathrm{Re})^2} \rightarrow$$

$$(m)g_x \propto \frac{(m)(Me)}{(Rx)^2} \rightarrow g_x = \frac{G(Me)}{(Rx)^2} \longrightarrow$$

$$\boxed{g_x = \frac{ge(\mathrm{Re})^2}{(Rx)^2}}$$

Centripetal force Equation:

$$\mathrm{CF} = \frac{(m)(v^2)}{R} \text{ where a} = \frac{(Vx)^2}{Rx} \text{ and a = g} \longrightarrow \boxed{g_x = \frac{(v_x)^2}{Rx}}$$

Combining these two equations gives the orbital equation:

$$\frac{(Vx)2}{Rx} = \frac{g_E)(\mathrm{Re})^2}{(Rx)^2} \rightarrow (Vx)^2 = \frac{(g_E)(Rx)(\mathrm{Re})^2}{(Rx)^2} = \frac{(g_E)(\mathrm{Re})^2}{(Rx)}$$

$$\boxed{Vx = \sqrt{(g_E)(\mathrm{Re})^2 / Rx}}$$

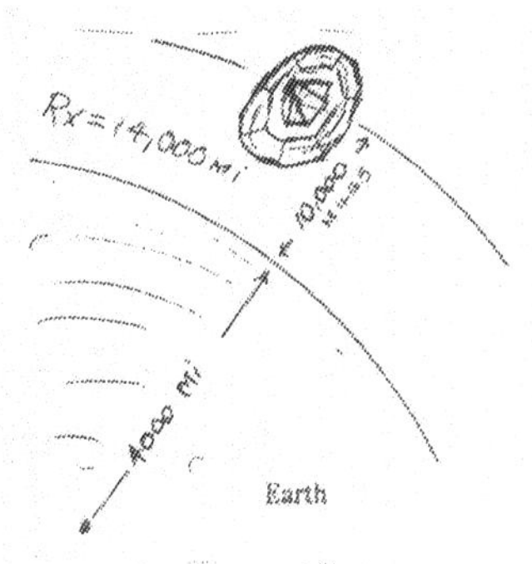

This velocity equation will calculate the required velocity to reach any orbit (x + R) from the center of earth. The rocket is fired vertically from the earth's surface and then is guided into a tangent path at the predetermined elevation. You will make several calculations of velocities at predetermined orbits.

Example (1) Calculate the velocity that is required to maintain the orbit of a space antenna that is 10,000 miles above the surface of the earth. (This is not the velocity needed to put it there, but only to keep it there.)

$$Vx = \mathbf{(RE)}\sqrt{(g_E)/(Rx)} =$$

$$=(4{,}000\text{mi})(5{,}2380 \text{ ft/mi})\sqrt{(32\,ft/s^2)/(14{,}000\;mi \; x \; 5{,}280\,ft/mi)} =$$

$$\mathrm{Vx} = 21.12 \times 10^6 \sqrt{(32)/73.92x10^6} = 13{,}900 \text{ ft/sec or } 9480 \text{ mi/hr.}$$

Your First Orbit

One of your class members will be placed in orbit 200 miles above earth. The orbiter is already up there, and by some magic, you will be placed in the orbiter and brought back later after the class has learned how to bring an orbiting craft back to earth??? At 200 miles, you will be able to see more of the detail on earth than at 10,000 miles for example. But, at 200 miles, you will be unable to tell whether or not there is life on earth. If you tune in on world communication, you will know there is a multiplicity of abundant life down there. That would be a great discovery for any person traveling from outer space. You will be up there at about the same elevation as Challenger when it does most of its space oriented research.

YOU ARE NOW IN ORBIT 200 MILES ABOVE EARTH. (It is important that you make the following calculations pertinent to your orbital location in space.) (1)Calculate your orbital velocity.

$$Vx = \sqrt{(g)(\mathrm{Re})^2/(Rx)} = \sqrt{\frac{(32f/s-s)(4000mix5{,}280f/mi)^2}{(4200mix5{,}280)}} = 29.02f/s-s$$

(2) Calculate gravity at 200 mi above earth.

$$Gx = (Ge)(\mathrm{Re}/Rx)^2 = (32f/s)\frac{(4000mi)^2}{(4200mi)^2} = 29.02f/s-s$$

(3) Calculate your centrifugal force given ship's weight is 4000 p. Do you feel the effect of CF in your capsule?

$$CF = (m)(v)2/(Rx) = \frac{(4000p)(25{,}370f/s)^2}{(4200mi)(5{,}280f/mi)} = 3628p$$

How do you like the feeling of being weightless? What factors make you weightless? Can you explain this to your class?

Already you have studied circular orbits and rotating objects. You are hardly any better off than the stone in David's slingshot. You know that your orbit is not circular but is elliptical. However, the orbit of your craft is near enough to that of a circle, that it can be treated as a full circle with your calculations having very small error.

(3) Calculate the circumference of your orbit about earth in miles and feet.

C=2(π)(Rx) = (6.28)(4200mi) = 26,376 mi=___________________ft.

This is the distance you travel each time you make one revolution about earth.

(4) Calculate your period of rotation about earth.

$$T = Period = \frac{Distance}{Velocity} = \frac{26{,}376mi}{17{,}298mi/hr} = 1.52hrs/rot$$

(5) Calculate the number of revolutions per day.

$$\text{Rev/Day} = \text{Frequency} = \frac{24hr/day}{1.52hr/day} = 15.79\ \text{Rev/Day}$$

(6) What part of a revolution do you make per hour?

$$\mathrm{Rev}/hr = \frac{15.79rev/day}{24hr/day} = .66rev/hr$$

(7) If your frequency of travel is .66 rev/hr, how many degrees of longitude does your orbiter pass through per hour of travel?

Angle(θ) = (360°/rev)(.66 rev/hr) = 236.8°

Rules Regulating Orbit

You are on the launch pad at Vandenberg AFB and are given the assignment to place two 400 kg objects in orbit, one at 500 km above earth and the other one at 1000 km above earth, You calculate the orbital velocity of the second object. The calculation for the first object is shown below.

$$V = \sqrt{(Ge)(\text{Re})^2 / (Rx)} = \sqrt{\frac{(9.9m/s-s)(6.36x10^6 m)^2}{(6.86x10^6 m)}} = 7600m/s$$

Why is the velocity at the higher distance less than that at the lower distance from the earth's center? Does gravity have anything to do with the forces that are acting upon the respective objects? How does this relate to the centripetal force acting upon each satellite at its respective point of orbit? Of course, the correct answers are that gravity is different for each point of orbit and that it is less for the more distant object than for the nearer object. For this reason, the respective centripetal forces are different, ie., CF for the farthest out object is less than that for the nearer object. For the same reasons, the orbital velocities required to keep an object in orbit at a given distance from earth decreases the farther out one gets form its center.

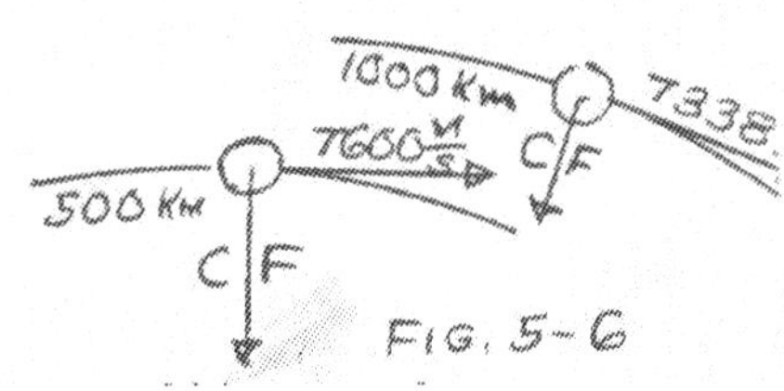

FIG. 5-6

Example (2) Calculate gravity 500 km and 1000 km from the earth's surface. Remember: Gravity and distance are inverse relationships.

$$(Gx)(Rx)^2 = (Ge)(\text{Re})^2 \rightarrow Gx = (9.8m/s-s)\frac{(6360km)^2}{(6860km)^2} = 8.43m/s-s$$

When you made the calculation for gravity at 1000 km, did you get 7.32 m/sec^2?

Example (3) Calculate the CF at 500 km and 1000 km above earth

$$CF = (m)(v)^2 / (Rx) = \frac{(400kg)(7600m/s)^2}{(6860km)} = 3{,}364Nt$$

What was your CF value for the 1000 km distance? Does the above conclusion about CF and distance hold true? Was your answer 2,740 N?

Exercise 18

Self Test

1. Write the general equation for an ellipse.
2. Write a formula that could be used to determine the length of each axis of an ellipse.
3. In the general equation of an ellipse, if a=19 and b=13, where would the foci be located on the major axis?
4. If a=7 and b=5, write the general equation of a given ellipse.
5. Calculate gravity at the following elevations above earth: (a) 175 miles; (b) 3200 miles.
6. Convert the above gravity calculations to mks units.

7. If gravity is as follows: (a) 27 ft/s-s; (b) 9.54 m/s-s. Calculate the distance form the surface of earth to each of these points in corresponding units.
8. The proposed International Space Station will orbit earth at 220 miles. Calculate (a) Its velocity of orbit; (b) Its period of orbit; (c) What will be gravity at the point of orbit?

Chapter VI

Rocket Powered Flight

Early Flight

Early man watched the flight of birds in wonderment. He marveled at the ease with which they flew and the vast distances they were able to cover. Cave drawings depicted their flight and some drawings were very accurate in showing wing design and winged flight characteristics. You need to remember that flight in outer space first began with flight in the atmosphere above earth. The notebooks of Leonardo da Vinci 1452–1519 contained detailed drawings of a two-part flapping wing. Otto Lilienthal of Germany made more than 2,000 glides in a single-winged plane. In some of these flights he glided more than 1000ft. On Dec. 17, 1903, Wilbur and Orville Wright made the first successful powered flight over sand dunes at Kitty Hawk, N.C. On the longest flight the craft was airborne for 59 seconds and traveled 852 ft. Powered flight was considered so important to modern warfare, that just 11 years after Kitty Hawk the British had four squadrons of planes. The history of the development of the powered airplane into modern times is well known by most students. If man was to get out into space, he would need another type of propulsion. What other developments were taking place that would make it possible for him to get beyond the limitations of the atmosphere?

The First Ballistic Missiles

You may be surprised to know that the Chinese developed the first rockets prior to 1232 A.D. As with the airplane, these skyrockets were dependent upon atmosphere for flight. Other than for festive occasions, most rockets were used for military purposes. An American physicist, Robert Goddard, was first to build and fire a liquid-fueled rocket in 1930. By 1936, the highest shot made by Goddard was 7,500 ft. At the same time, a German army research group headed by Walter Dornberger and Wernher von Braun, had fired rockets to 400 ft. Wernher von Braun is considered to be the father of the modern rocket, and is given credit for the development of the V-2, which is the worlds first large liquid-fuel ballistic missile. This rocket stood 46 ft. high and 66 inches in diameter. It weighed 27,000 pounds, and generated a thrust of 56,000 pounds. It achieved an altitude of 65 miles and a range of 200 miles with a payload of 2000 pounds.

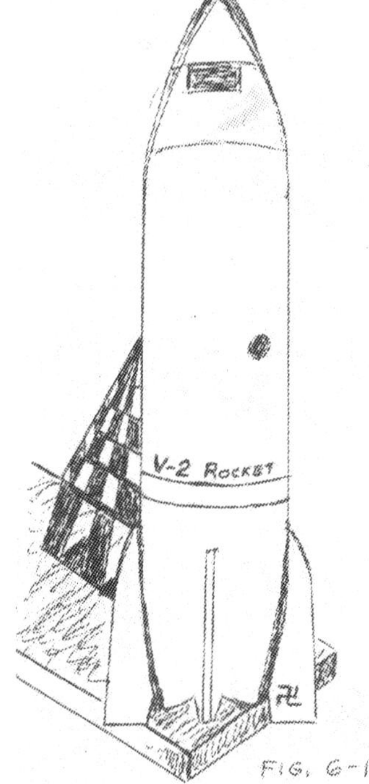

FIG. 6-1

Rocket Propulsion

Multi-stage rockets are the rockets used for space travel, space research and exploration, and for delivering large nuclear warheads. Single stage rockets are principally used by the military to deliver relatively small explosive charges on the battle field. However, some single stage rockets can be used to orbit small satellites into some of the lower orbits.

The propulsion of these rockets requires fuel that burns with great energy over short periods of time. The propelling energy is more like a controlled explosion than a simple combustion. For many years, up until present times, the effective way to produce the forces needed to propel rockets over large distances, was to mix chemical liquids in a combustion chamber and to ignite them. Even these fuels had their limitations, but did the job. Only in recent times, has science produced solid propellant engines that could thrust large payloads into space. The great advantage of solid propellants is that they do not tend to be as corrosive as liquids, do not leak out of the engine and can be stored more easily for long periods of time.

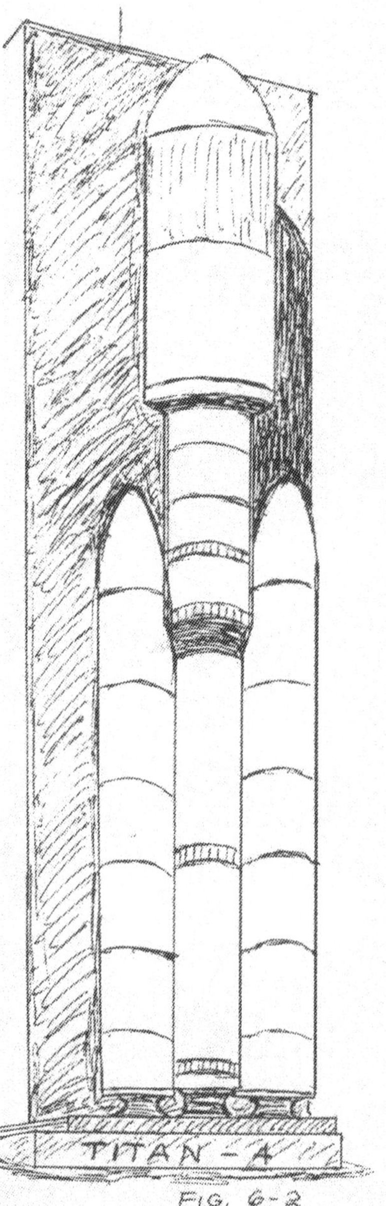

FIG. 6-2

Types of Rocket Engines

Multi-stage or "step) rockets, as they are sometimes called, have the advantage of being able to use both types of fuel, each in different stages of the rocket being fired. The main reason for using multi-stage rockets is that once each stage has fired, it can be dropped away from the main vehicle which becomes immensely lighter. As each section of the rocket is fired, the remaining stages then add to the velocity of the first, each remaining stage going faster and faster.

The Titan 4 rocket shown to the right is a multi-stage rocket that uses both liquid and solid propellants. It is 204 feet long and 16 feet in diameter. The two first stage solid boosters have a combined 3.2 million pounds of thrust. The first stage liquid fueled engine adds 546,000 pounds of thrust. After all three first stage engines complete their burn. The second stage ignites providing 104,000 pounds of thrust. The final third stage then adds an additional 33,000 pounds of thrust to complete the burn period.

Liquid Propelled Engines

Liquid fueled engines consist of two tanks and a mixing chamber in which the liquids combine to combust into gases which then exit at high velocity from the thrust chamber. One tank holds the primary fuel which can be gasoline, alcohol, aniline, kerosene, etc. The second tank hold the oxidizer which can be nitric acid, LOX (O), nitrogen tetroxide, etc. The fuel and oxidizer are then pumped into the mixing chamber under pressure.

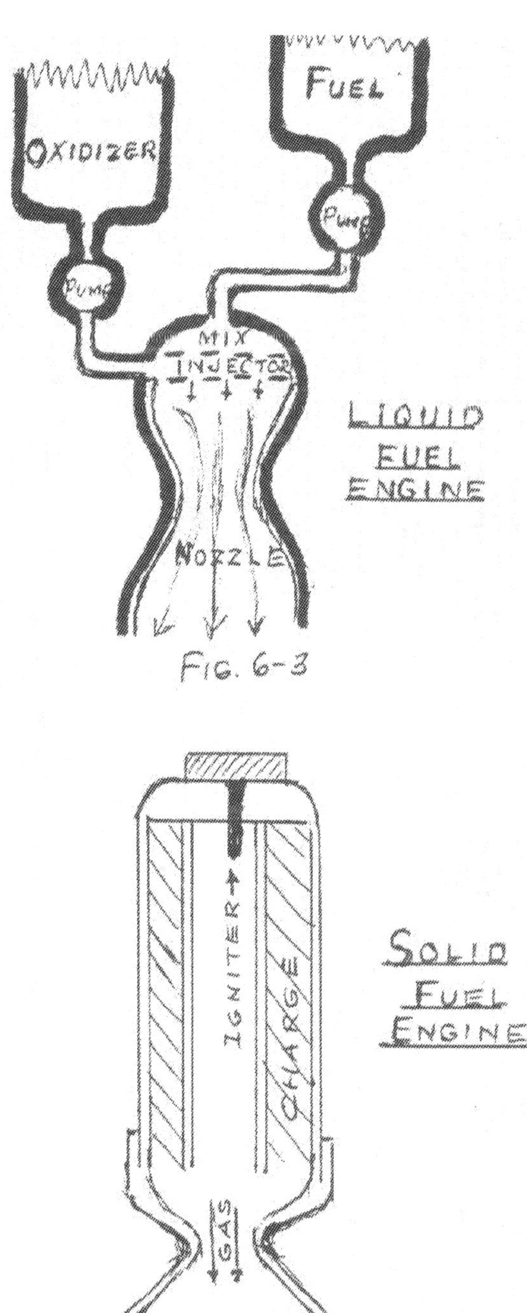

FIG. 6-3

FIG. 6-4

Solid Propellant Engines

In solid propellants, the fuel and the oxidizer are combined in the solid substance. The solid substance is usually a powdery or rubbery mixture which then burns in the same chamber in which it is stored. For this reason, the casing of the solid propellant is both tank and combustion chamber. It took many years of testing and development of solid propellant engines before they could be successfully fired. All engines develop great internal vibration. The problem with the original solid-type engines was that the fuel base was so brittle that it cracked and exploded. It was not until latex substances were used to form the fuel that the first successful firings took place.

Improving Engine Efficiency

The design of the exit chamber through which combusted gases flow is of primary importance. In general, the best design results when the expanding portion of the exhaust chamber is three to four times as long as the constricting part of the chamber.

Assuming the design of the chamber is optimum, there are two other ways to get more velocity out of the exhaust gases. (a) The first is by increasing the energy output of the propellant, ie., chemical (b) The second is by using light weight materials in the rocket.

Propulsion Vocabulary

In order to become knowledgeable about rocket propulsion, you must learn several new terms. These terms are as follows: THRUST, SPECIFIC IMPULSE, EXHAUST VELOCITY, THRUST/WT(total), WT(total)/WT(fuel) and WT(total)/WT(burnout).

THRUST (the measure of rocket engine performance is the amount of reaction force (push) created in the process of combustion and ejection of gases. As hot gases are produced in the combustion chamber, they expand and begin their exit through the nozzle of the rocket engine at less than the speed of sound. As the gases approach, and begin to pass through the narrow exit chamber, they speed up to the speed of sound and then accelerate to higher and higher velocities. The reaction of the rocket to these expanding gases is to move away from them with greater and greater velocity. In order for the rocket to move upward, the thrust in pounds, must be greater than the weight of the rocket. This is the 'thrust to total weight of the rocket) ratio.

SPECIFIC IMPULSE (Another measure of engine performance) is defined as the amount of thrust which can be derived from each pound of propellant in one second of engine operation. The formula:

$$\text{Specific Impulse} = \frac{Thrust(F)xTime(t)}{\Pr opellant\_weight(W)}$$

EXHAUST VELOCITY is the speed in ft/sec and m/sec, at which ignited gases are expelled from the engine. (Exhaust velocity is equivalent to specific impulse as an index of operation performance.) Formula:

$$\text{Exhaust Velocity(v)}=\frac{Thrust(F)xGravity(g)xTime(t)}{Weight\_of\_fuel(W)}$$

MASS RATIO is the quotient of the total weight of the rocket before firing divided by the weight of the fuel. The formula:

$$R = \frac{WT(total)}{WT(fuel)}$$

BURN RATIO is the quotient of the total weight of the rocket before firing divided by the weight of the rocket after burnout. The Formula:

$$R = \frac{WT(total)}{WT(total) - WT(fuel)}$$

The V-2 the Proto-type of the Modern Rocket

The Father of the modern rocket had its beginning near the end of WW II. Had the invention of this rocket been one year earlier, the Allied countries might not have won the war. This rocket was capable of delivering a one ton explosive two hundred miles away and reaching a height of 60 miles. The V-2, a one stage rocket, provides a simple way to illustrate the information on the previous page. The V-2 data provided in the drawing below is essentially correct.

(1) Determine the R(f), THRUST TO WT. RATIO of the V-2 rocket, given the engine thrust to be 56,000 p and the WT(tot) to be 29,000 p.

R(f) = Thrust / W(tot) = F / W(tot) = 56,000p / 29,000p = 1.931

(2) Calculate the rocket's SPECIFIC IMPULSE (I) in seconds, given the burn time to be 85 seconds.

SPECIFIC IMPULSE = (F)(T) / W(fuel) = (56,000p)(85s) / 19,040p = 250s

(3) Calculate the EXHAUST VELOCITY (Vx) = (I)(G) = (250s)(32 f/s-s) = 8000 f/s (Remember: This is not the velocity of the rocket, but is the velocity of burning gases escaping through the engine opening into space.)

(4) Calculate the BURN RATIO (Rburn) = W(tot) / [(Wtot) – W(fuel)] = 29,000p/9960p. R(burn) = 2.91

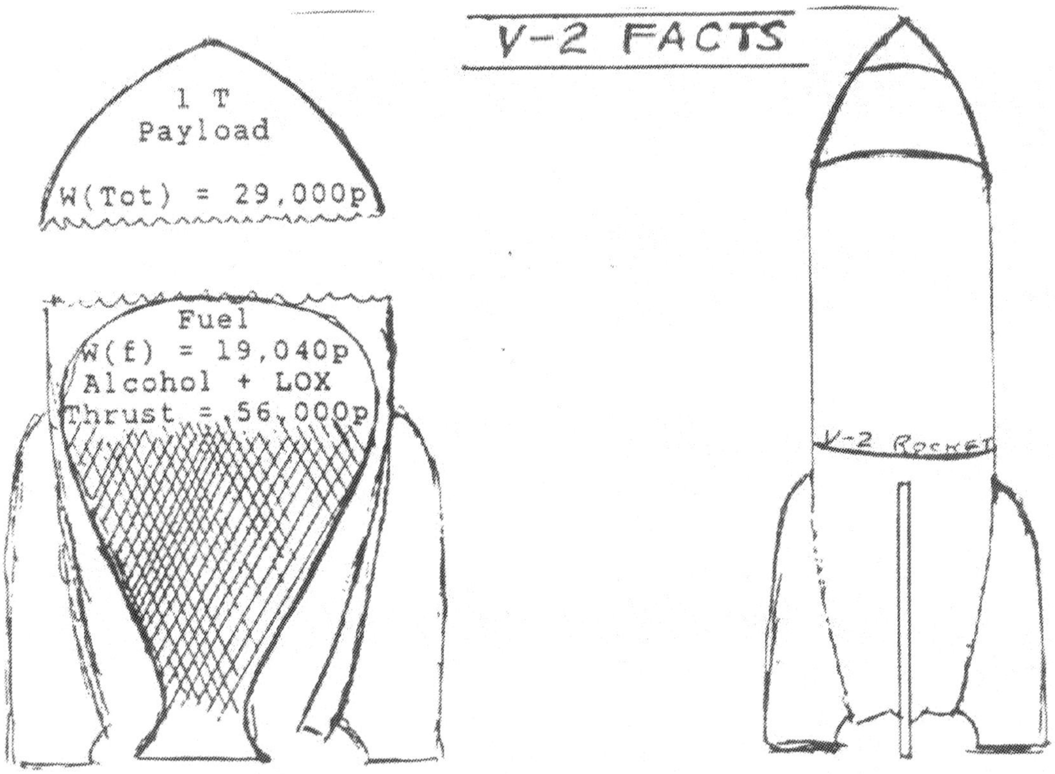

V-2 FACTS

1 T
Payload
W(Tot) = 29,000p

Fuel
W(f) = 19,040p
Alcohol + LOX
Thrust = 56,000p

V-2 Rocket

Exercise 19

Self Test

1. The father of the modern rocket was _____.
2. The first person to use liquid fuel in rockets was _____.
3. The first sustained flight was done by whom?
4. The V-2 rocket was the first_____ _____.
5. The two principal components of liquid fueled engines are _____ and _____.
6. The principal supporting compound for solid fuels was _____.
7. Name two kinds of oxidizers and two kinds of fuels used in liquid-type engines.
8. The exit chamber of the engine nozzle should be _____ to _____ times as long as the constricting chamber.
9. After passing through the narrow portion of the nozzle, the gases continue to _____ and to increase in _____.
10. Name two ways of increasing the burning efficiency of fuel.
11. Define thrust and name its principal unit.
12. Define specific impulse and name its principal unit.
13. Calculate the thrust to weight ratio of an engine if its thrust is 800,000 pounds and the weight of its fuel is 420,000p.
14. What is the specific impulse of the rocket engine in problem (13) if the fuel burns for 150 seconds?
15. Calculate the velocity of gas flow from the engine in problems (14).

Second in Orbit

One other of you will be sent into orbit before the first one gets back to earth. It is necessary to develop the physics of sending a rocket into orbit before attempting to bring one down from orbit. You will decide which is easier, going up or coming down. We will use a Vanguard-type-rocket with LOX and Kerosene as the principal oxidizer and fuel. The specific impulse of this fuel at sea level is 250 s. But first things first. You need to consider the mass ratio of each successive burn stage of the rocket, before you begin thrust and velocity calculations.

The drawing below is that of a large three stage rocket as it moves through successive stages in which it burns its fuel in each stage and then drops that stage away to begin the next burn and stage reduction. The rocket weights 1,411,000 pounds on the pad with fuel and payload. Stage weight with fuel weight are given. You are to calculate the rocket to fuel weight ratio and the rocket to burnout weight ratio for each stage.

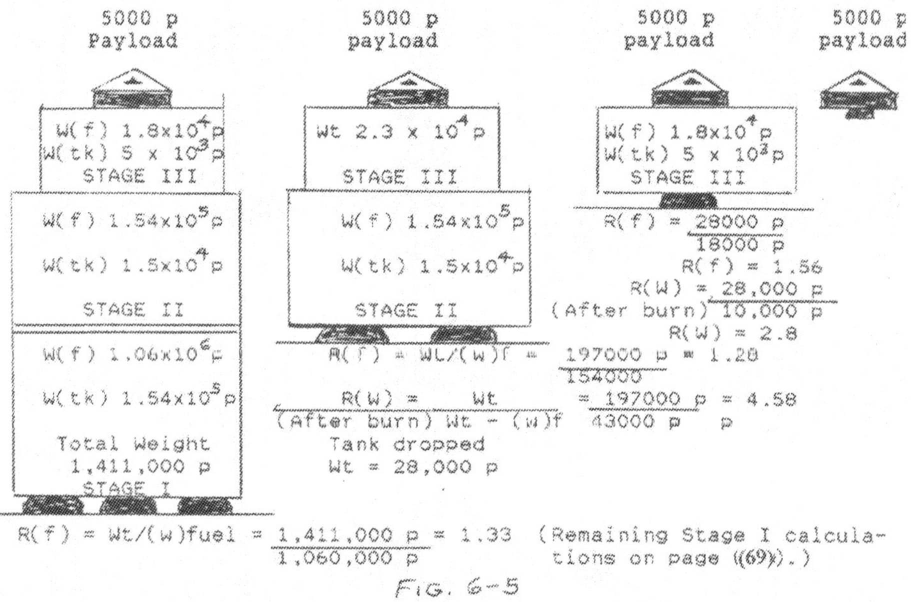

FIG. 6-5

Stage I Calculations Continued:
After burn weight = Wt – (w)fuel = 1,411,000 P – 1,060,000 P = After burn weight = 351,000 P

$$R(W) = Wt/(after\ burn\ wt) = \frac{1{,}411{,}000p}{351{,}000p} = 4.02$$

The ratio of (total stage weight) to (after burn weight) is most important because it provides a way for you to calculate the final increase in the velocity of the rocket due to weight reduction that results from the burn of fuel in each stage. In all of your previous calculations of acceleration and velocity, you worked with dynamic objects of constant mass. A rocket goes faster and faster because of mass reduction due to converting mass into gases that exit from the rocket engine. After you have completed the next exercise, which is similar to that above, you will then use this information to place a manned rocket in orbit.

Exercise 20

Self Test

The rocket in this exercise is similar to the one just shown, but has different "stage" information given. You are to calculate the R(f) or (mass ratio) and the R(W) or (weight-burn ratio) for each stage of the rocket. Assume that each stage is dropped form the rocket.

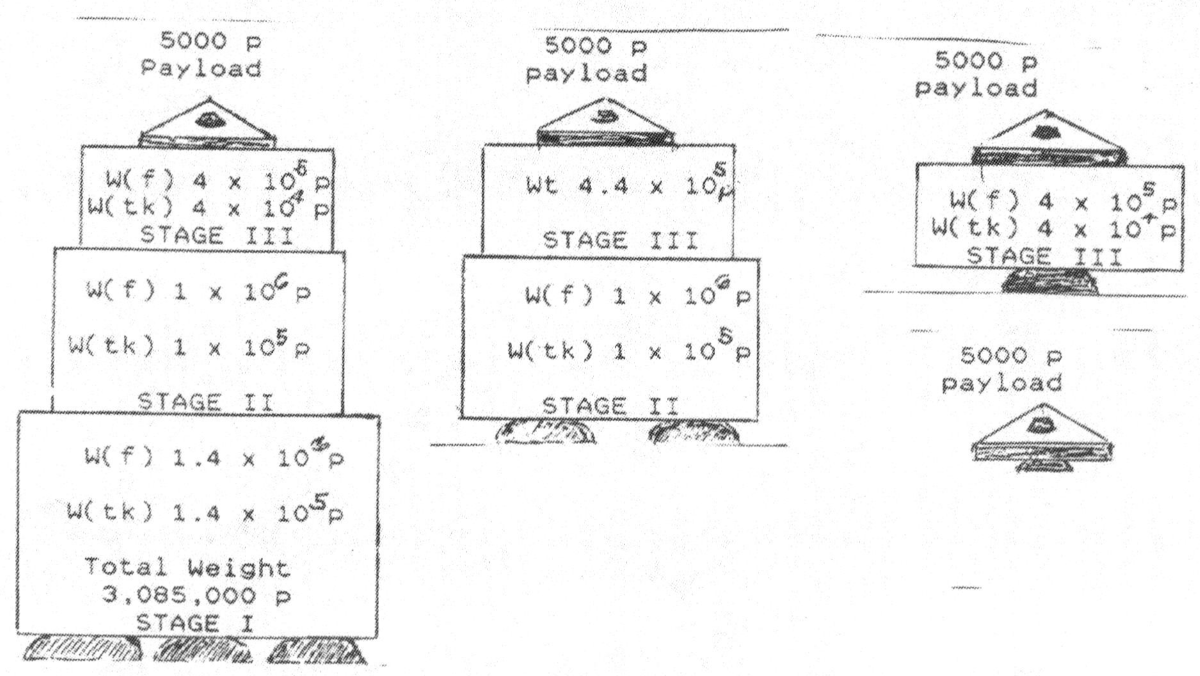

<u>First Multi-stage Rocket to Orbit</u>

The rocket used in the example on page (69) is capable of placing a payload of 5000 p into orbit. The purpose of this flight will be to place two class members in orbit at 320 miles above earth. You will now begin to learn how to do calculations required to do this. The fundamental rocket chosen is the one proposed on page (69). The orbital craft must be given a final velocity of 25,015 ft/sec in order to stay in orbit at 320 miles above earth. There will be some fuel left in the third stage of the rocket in order to complete the "adaptation to orbit" maneuver, ie., to round out the orbit.

Pre-data
Rx = 300 mi.
Vorb = 25,073 ft/sec
Stage I Data

F=2,000,000p
G=32 ft/s—s
Wf=1,060,000p
T=133 sec
Vg=8030 ft/sec
R=4.02
θ=40°
I-250 sec
Wb=351,000p
Wtk=154,000p
Vi=7910 ft/sec

STAGE I (Calculations)

$$\text{Vgas}=\frac{(Thrust)(Gravity)(time)}{Fuel\_weight}=\frac{(F)(g)(T)}{(Wf)}=$$

$$\text{Vgas}=\frac{(2{,}000{,}000p)(32ft/s-s)(133s)}{1{,}060{,}000p}=$$

Vg=8030 ft/sec

Given R=4.02(calc. On page (58) log 4.02=1.391

Vi=(Vg)(log R)-(g)(T)(Cosθ)=

Vi=(8030 ft/sec)(1.391) – (32 ft/s-s)(133s)(Cos 40°)=

Vi=11,170 ft/sec-3260 ft/sec=7910 ft/sec

The earth is rotating at 1534 ft/sec. This velocity effects only horizontal component of velocity and will be added later.

Stage II Data
Wt=197,000p
Wf=154,000p
T=154 sec
F=300,000p
Vg=9600 ft/sec
Wtk=15,000p
R=4.58
Wb=43,000p
θ=60°
g'=30.82 ft/s-s
Vii=12219 ft/sec
Vtot=20,129 f/s
Iii=300sec

STAGE II (Calculations)

$$Vg=\frac{(F)(g)(T)}{Wf}=\frac{(300{,}000p)(32ft/s-s)(154s)}{154{,}000p}=9600 \text{ ft/s}$$

Given R=4.58 ⟶ ln 4.58 = 1.52

Vii-(Vg)(ln R) = (g')(T)(Cosθ)=

Vii = (9600 ft/s)(1.52) – (30.82 ft/s-s)(154s)(Cos 60°)=

Vii=14592 ft/sec = 2373 ft/sec = 12219 ft/sec

Total Velocity = Vi + Vii = 7910 ft/sec + 12219 ft/sec =

Vtotal = 20,129 ft/sec

This velocity is only 4886 ft/sec short of the required velocity.

You can see that you are only 4886 ft/sec short of the required velocity to enter orbit. It will not be necessary to use all of the fuel in the third stage tank, as 10,000 p will bring you up to orbital speed. There will be enough fuel left to make any orbital correction required for adaptation to orbit.

Stage III Data
F=22,000p
Wt=28,000p
Wf=10,000p
T=127sec
θ=80 degrees
I = 280 sec
Vg = 8,941 f/s
Viii=3,334 f/s
Wb=18,000p
G" = 28.04 f/s-s

STAGE III (Calculations)

$$Vg=\frac{FxGxT}{Wf}=\frac{(22{,}000p)(32f/s-s)(127s)}{10{,}000p}$$

Vg=8941 ft/sec

Given R = 1.556 ⟶ In(1.556) = .442

Viii = (Vg)(In R) = (G)(T)(Cos θ) =

Viii = (8941f/s)(.442) – (28.04 f/s)(127s)(Cos 80°)

Viii = 3952 f/s – 618 f/s – 3334 f/s

Calculate your final orbital velocity which now includes the velocity of the earth's rotation.

X(v)orbit = V(R) + Viii + Ve = 20129 f/s + 3334 f/s + 1534 f/s =

X(V) orbit = 24997 f/s

The final velocity of your ship is 18 ft/s less than the required velocity to achieve orbit. Such a small difference is easily corrected. Note that 1534 f/s, the earths velocity of rotation, is imparted to the horizontal component of your ship's velocity.

There are 8,000 p of fuel left in the third stage tank. A small amount of this fuel will be used to increase the ship's velocity to the orbital velocity. BUT, did your ship reach an elevation of 300 miles? NEXT, you need to investigate the calculations of orbital height and distance. There is still time left to make all of the required adjustments to orbit.

If you check back over the calculations, you will see that your first stage motion was at an average angle of 40 degrees. The starting motion, just off of the pad was at 0 degrees with the vertical. As the ship moved farther and farther out from earth, it turned more and more away from the vertical. During the second stage, the average angle of the rocket was 60 degrees. The rocket is assumed to be moving in an arc away from earth, with the final stage bringing it parallel to earth. The average angle with the vertical during the third stage is 80 degrees. The rocket finally reaches orbital velocity as it's angle becomes 90 degrees. At this time it is traveling parallel to the surface of the earth below.

Stages I-III Calculations of Vx,y and Sx,y results. (S = distance)

STAGE I

The first vector drawing represents vertical and horizontal velocities of the rocket. The second vector drawing depicts vertical and horizontal distances.

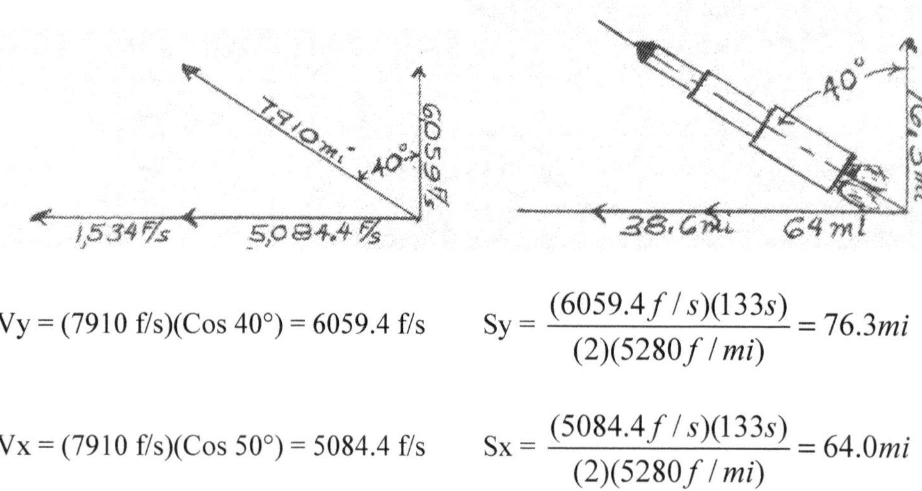

Vy = (7910 f/s)(Cos 40°) = 6059.4 f/s $\qquad$ $Sy = \dfrac{(6059.4f/s)(133s)}{(2)(5280f/mi)} = 76.3mi$

Vx = (7910 f/s)(Cos 50°) = 5084.4 f/s $\qquad$ $Sx = \dfrac{(5084.4f/s)(133s)}{(2)(5280f/mi)} = 64.0mi$

Add to the horizontal distance the distance the rocket moves due to velocity imparted by the rotation of the earth.

Sx'=(1534 f/s)(133 s)/(5,289 f/mi) = 38.6 mi.

STAGE II

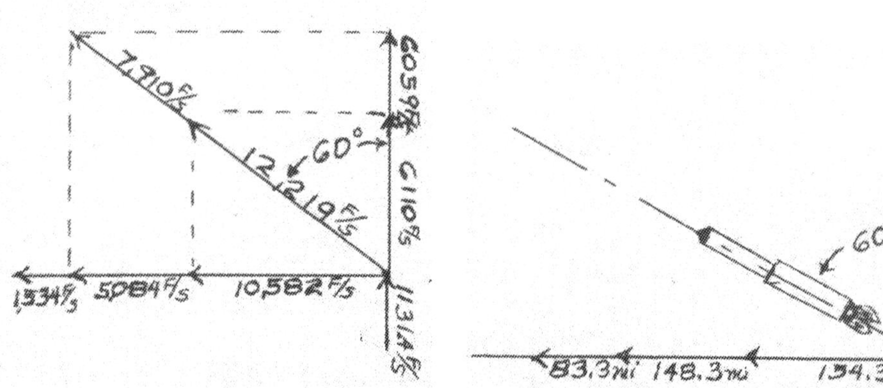

iiVy-(Sin 30°)(12,219 f/s) = 6110 f/s
iiVx+(Cos 30°)(12219 f/s) = 10,582 f/s
ΔVy = 6059.4 f/s-gt =
ΔVy = 6059.4 f/s – (30.82 f/s-s)(154 s)
ΔVy = 1131.4 f/s (remaining vertical (V) component)

$$iiSy = \frac{(.5)(6110f/s)(154s)}{5280f/mi} = 89.1mi$$

$$iiSx = \frac{(.5)(10,582f/s)(154s)}{5280f/mi} = 154.3$$

$$iSx = \frac{(5084.4f/s)(154s)}{5280f/mi} = 148.3mi$$

$$eSx = \frac{(1534f/s)(154s)}{5280f/mi} = 44.7mi$$

ΔSy = VT-.5(g)(T) =

ΔSy = ((6059.4 f/s)(154 s) - .5(30.82 f/s-s)(154 s))/(5280 f/mi) = 107.5 mi
Sy(tot) = 76.3 mi + 89.1 mi + 107.5 mi = 272.9 mi
Sx(tot) = 64.0 mi + 38.6 mi + 148.3 mi + 44.7 mi = 295.6 mi

STAGE III CALCULATIONS

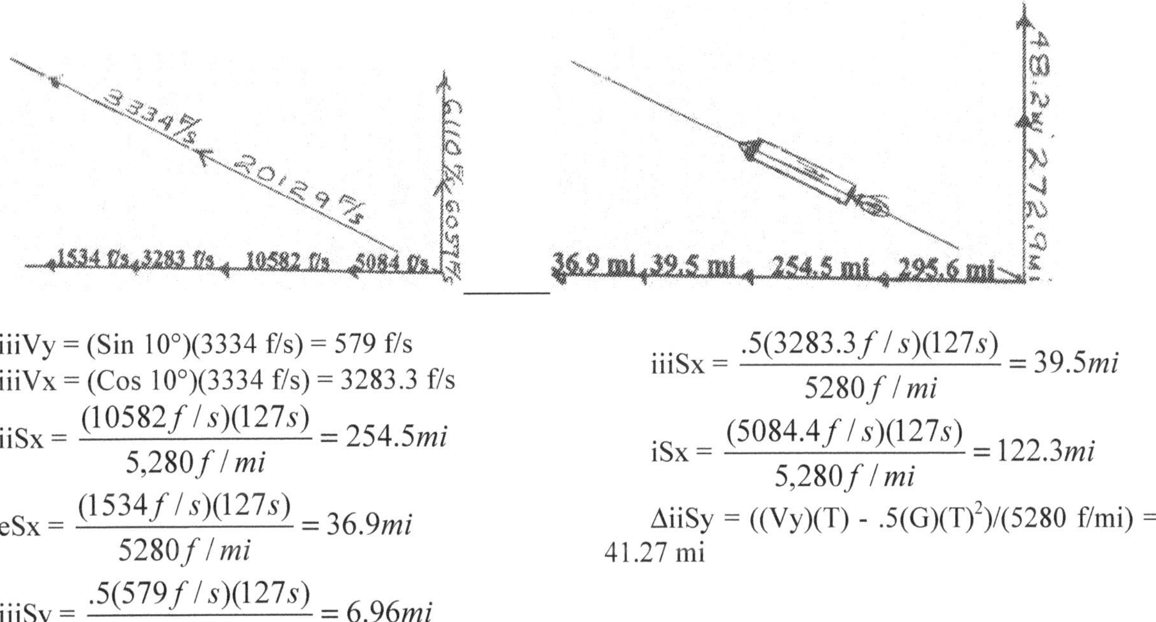

iiiVy = (Sin 10°)(3334 f/s) = 579 f/s
iiiVx = (Cos 10°)(3334 f/s) = 3283.3 f/s

iiSx = $\dfrac{(10582 f/s)(127 s)}{5{,}280 f/mi} = 254.5 mi$

eSx = $\dfrac{(1534 f/s)(127 s)}{5280 f/mi} = 36.9 mi$

iiiSy = $\dfrac{.5(579 f/s)(127 s)}{5280 f/mi} = 6.96 mi$

iiiSx = $\dfrac{.5(3283.3 f/s)(127 s)}{5280 f/mi} = 39.5 mi$

iSx = $\dfrac{(5084.4 f/s)(127 s)}{5{,}280 f/mi} = 122.3 mi$

ΔiiSy = ((Vy)(T) - .5(G)(T)^2)/(5280 f/mi) = 41.27 mi

ΔiiSy = (20,129 f/s)(127 s) (Cos 80°) - .5(28.04 f/s-s)(127 s)^2)/(5280 f/mi) = 41.27mi
What total horizontal distance did your orbiter travel?
Total Sx = (iSx + iiSx) + iiiSx = 295.6 mi + 330.9 mi = 626.5 mi
What total height did your orbiter reach?
Total Sy = (iSy + iiSy_ + iiiSy = 272.9 mi + 48.23 mi = 321.13 mi

Congratulation: you have brought your ship within 1.13 miles of correct orbit and you are only 18 f/s from correct orbital speed. This is a job well done.

A Basic Program Check

This will be your first introduction to Basic Programming. There is an introductory chapter on this subject beginning on page (84). Turn the page and you will see a Basic program with lots of data. Basic is the original computer language used to develop PC computers. You will soon prove to yourself that you can use equations to solve complex rocket problems, but you are about to discover another, more accurate way to solve these problems. This Basic program permits you to use the same equations that you used before, but to make many consecutive calculations beginning with zero, progressing through a series of steps until you have solved the equation 45 times and recorded all of the data for each solution. There are some space problems that can be solved only in this manner if you have not had higher mathematics.

This Basic Program is used as a check on our calculations on page (70). Notice that the final velocity result in the Basic chart is 7993.62 ft/sec, but that the final result on our calculations using rocket equations was 7,910 ft/sec. If you subtract these values, you will determine the numerical error to be 83.62 ft/sec. This amounts to an error of only 1% which is really a small error. These results are good enough to place you in orbit and to safely return you back to earth from orbit. Take note of the fact that 7,910 ft/sec is a linear result, whereas 7993.62 ft/sec is a curvilinear result, ie., one is straight line and the other curved.

YOUR FIRST BASIC PROGRAM

This Basic Program will solve the First Stage Rocket problem on Page (70). The data will be used in Exercise (21).

FIRST IN ORBIT – STATE I = firstorb
X = FUEL WT (Pounds)
F = THRUST (Pounds)
I = IMPULSE (Pounds)
G = AVERAGE GRAVITY (FT/sec-sec)
C = VELOCITY OF GAS FLOW (Ft/sec)
V = VELOCITY OF THE SHIP (Ft/sec)
W = PRE-BURN WT OF SHIP (Pounds)
A = ANGLE (Degrees)
T = TIME OF BURN (Seconds)
N = FUEL WT AFTER PER SECOND BURN (Pounds)

V(Velocity)	T(Time)	A(Angle)	ΔX(Fuel Wt)	R(Burn Ratio)
0	0	0	1411000	0
44.77437	3	.9090001	1386910	1.722034E-02
92.03432	6	1.818	1362820	.0347426
141.9337	9	2.727	1338730	5.257725E-02
194.6336	12	3.636	1314640	7.073568E-02
250.3015	15	4.545	1290550	8.923008E-02
309.1078	18	5.454	1266460	.108073
371.2288	21	6.363	1242370	.1272777
436.848	24	7.272001	1218280	.1468586
506.1544	27	8.181	1194190	.1668305
579.3441	30	9.09	1170100	.1872093
656.6235	33	9.999001	1146010	.2080123
738.2034	36	10.908	1121920	.2292571
824.3058	39	11.817	1097830	.2509631
915.1628	42	12.726	1073740	.2731507
1011.017	45	13.635	1049650	.2958419
1112.122	48	14.544	1025560	.3190599
1218.744	51	15.453	1001470	.3428297
1331.165	65	16.362	977380	.3671784
1449.683	57	17.271	953290	.3921348
1574.611	60	18.18	929200	.4177299
1706.282	63	19.089	905110	.4439975
1845.05	66	19.998	881020	.4709736
1991.293	69	20.907	856930	.4986977
2145.417	72	21.816	832840	.5272124
2307.852	75	22.725	808750	.5565641
2479.068	78	23.634	784660	.5868035
2659.567	81	24.543	760570	.6179858
2849.894	84	25.452	736480	.6501718
3050.645	87	26.361	712390	.6834285
3262.464	90	27.27	688300	.7178291
3486.064	93	28.179	664210	.7534555
3722.227	96	29.088	540120	.7903983
3971.815	99	29.997	616030	.8287584
4235.787	102	30.906	591940	.8686486
4515.219	105	31.815	456850	.9101966
4811.311	108	32.724	543760	.9535459
5125.422	111	33.633	519670	.9988599
5459.097	114	34.542	495580	1.046325
5814.1	117	35.451	471490	1.096156
6192.47	120	36.36	447400	1.148601
6596.569	123	37.269	423310	1.203949
7029.166	126	38.178	399220	1.262541
7493.542	129	39.087	375130	1.324781
7993.517	132	39.996	351040	1.391154

```
5 LPRINT "FIRST IN ORBIT – STATE I – firstorb"
10 LPRINT "X = FUEL WT (Pounds)"
20 LPRINT "F = THRUST (Pounds)"
30 LPRINT "I = IMPULSE (Pounds)"
40 LPRINT "G = AVERAGE GRAVITY (Ft/sec-sec)"
50 LPRINT "C = VELOCITY OF GAS FLOW (Ft/sec)"
60 LPRINT "V = VELOCITY OF THE SHIP (Ft/sec)"
70 LPRINT "W = PRE-BURN WT OF SHIP (Pounds)"
80 LPRINT "A = ANGLE (Degrees)"
90 LPRINT "T – TIME OF BURN (Seconds)"
100 LPRINT "N = FUEL WT AFTER PER SECOND BURN(Pounds)
110 X = 1060000!
120 F = 2000000!
130 I = 250
140 G = 31
145 R = 0
150 A = 0
155 L = 0
160 W = 1411000!
165 N = 0
170 T = 0
180 C = 8000
185 LPRINT "V(Velocity) T(Time)  A(Angle) ΔX(Fuel Wt)"
187 LPRINT 0,0,0,14110001,0
190 T=T+3
195 D = 8030*T
200 N = 14110001 – D
205 L = W/N
210 R = LOG (L)
220 A = .303*T
230 V = 8000*R – (COS(A/57.32)*G*T)
240 IF T = 135 THEN GOTO 270
250 LPRINT V,T,A,N,R
260 GOTO 190
270 STOP
```

Exercise 21

Self Test

Use the results of the Basic Program to carry out this exercise. Observe that the data is underlined every third row. Use only the data above each line for your charts and graphs.

1. Make up a data chart which includes Velocity, Time and Angle.
2. Graph V vs T using the vertical axis for your V value. (Your graph should be similar to the graph drawn to the right, but should not be the same as the graph.) Draw a straight line from the origin to the highest point on the graph and drop a vertical line to the horizontal axis. Note that this forms a right triangle. Calculate the area of the triangle. Do you see that this area is the distance the rocket traveled in the first stage? How far is this in miles?

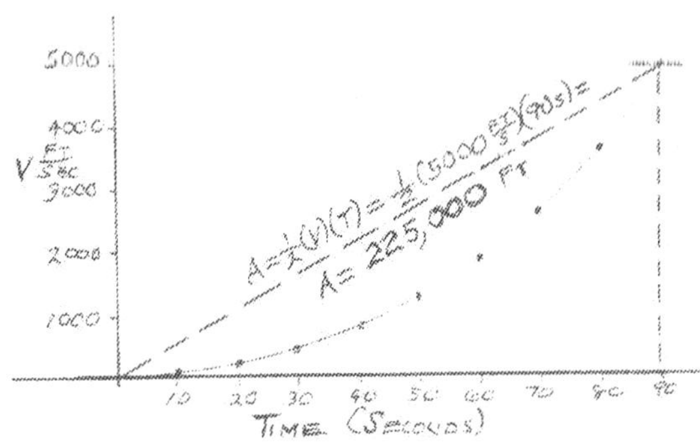

3. Show the calculation of this area and the check calculation using the final velocity in your chart. These two calculations should be the same.

A second graph, nearly identical to the first graph is drawn to the right. This time the diagonal line is drawn to split the difference in area that is due to the arc of the curve. You may draw this line on your original graph. Do you see that if the area under the curve is the true distance the rocket traversed in Stage I, then the arc of the curve must be taken into consideration if the distance is correct. If your graph is big enough, and if you are careful when you draw the line, the resulting distance will be the correct distance the rocket flew. (4) Show the calculation of the true area under the curve. (b) What is this distance in miles? (5) By how much was your result different form the original calculation? This result is nearly the true distance the rocket traveled in Stage I. (6) Using the final angle of 40°, calculate the vertical and horizontal distances the ship traveled in feet and miles.

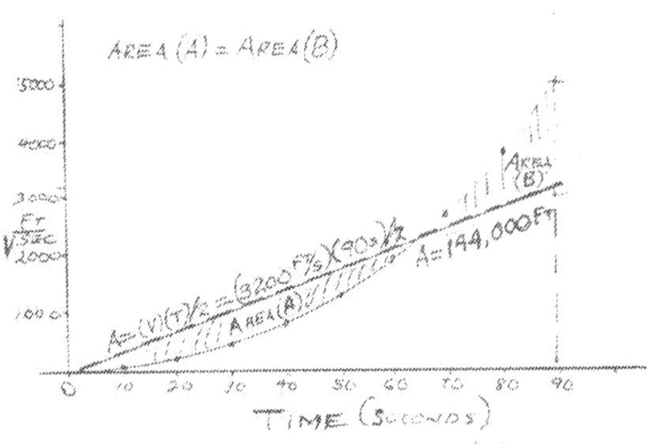

Exercise 22

Self Test

In solving the following problems, use the sequence of operations as follows: equation, substitution (necessary substitution steps), and answer, as your means to the solution of the problem.

(1) Given: the time of burn for a single stage rocket as 155 seconds; the thrust as 8000 pounds; the weight of fuel as 6000 pounds. Calculate the exhaust velocity of the rocket.
(2) If the burn ratio (R) is 3.5, calculate the final velocity of the rocket at burnout. Given: the average values for (g) and (θ) as 31 ft/sec-sec and 55 degrees.
(3) Use the final velocity calculation form problem (2) to determine the altitude of the rocket at final burn in ft/sec and mi/hr.
(4) Calculate gravity at this altitude and compare this with the average gravity you were given in problem (2). By how much does your answer differ from 31 ft/sec-sec?
(5) Assume the rocket traveled in a quarter circular arc from earth to its burnout location. How far in feet and miles did it travel?
(6a)Calculate the velocity of earth rotation in mi/hr and in ft/sec. (b) Taking this into consideration, what is the final resultant velocity of the rocket at burnout? (c) Relative to earth, what is the velocity of the rocket?
(7) Use the burn ratio given in problem (2) to calculate the weight of the rocket (W) prior to burn.

Return From First Orbit

Now that you have successfully explored the mathematics of placing an object in orbit, it is time to bring down your fellow classmate from his orbit 200 miles above the earth. If you have climbed a cliff or a high tree, you know it is harder to come back down than it was to go up. It is easy enough to bring an object out of orbit and drop it to earth, even as it is easy to fall from the cliff or tree. In all three instances, you may be "dead on arrival." Most shooting stars (falling meteors) burn up in the atmosphere and do not reach earth. The problem is to bring down the orbiter and its occupant safely. The orbiter will come down very fast, but not so fast that it will overheat excessively. The outer surface is coated with a ceramic material that peals away carrying the heat with it. This gives great protection to the occupant of the craft. The orbiter must be kept in the same orientation during the entire trip back to earth. Should it begin to tumble, the threat to inside life becomes imminent. Near the end of the flight, the drogues will be deployed to orient and slow the craft, then three parachutes will open to further slow down the craft for splash-down. The orbiter and class astronaut will be landing in the Atlantic Ocean just off the coast at Cape Canaveral.

Your ship is moving in the direction of the earth's rotation with a velocity of 25,370 ft/sec. You must reverse maneuver the ship, so that the retro-rockets can be fired, in order to slow it down. Your orbital craft weights 2.5 tons, which includes 2,000 p of fuel with a thrust capability of 30,000 p. The engine will consume all of the fuel during 30 seconds of burn. Calculations are as follows:

$$\text{Vgas} = \frac{30{,}000px32f/s - sx30s}{2{,}000p} = 14{,}400f/s$$

$$R = \frac{Wt(total)}{w(burn)} = \frac{5000p}{3000p} = 1.67 \rightarrow \ln(1.67) = .51$$

V(ship) = (Vg)(ln(R)) = (14,400 f/s)(.51) = 7384 f/s

The resultant velocity of the ship is:

V(R) = 25,370f/s – 7384f/s = 17985 f/s

Upon retro-fire, the ship will begin to fall toward earth at an increasingly faster rate of descent. Why? Because now gravity is a greater force than is the centrifugal force. What is the magnitude of this acceleration downward? Recall the CF formula. The bracketed part of the formula is the centrifugal acceleration.

CF = (m)[$(v)^2$/(Rx) from which [$(v)^2$/(Rx)] = acceleration = Acf

---

Acf = $(17{,}985\text{f/s})^2$/(4,200mi x 5,280f/mi) = 14.59 f/s-s

---

ΔGravity = Gx – Acf = 29.02f/s-s – 14.59f/s-s = 14.43 f/s-s

Return Trip Calculations Continued

The ship begins its descent at the instant your retro-rockets fire. The reduction in velocity immediately changes the condition of balanced forces, ie., gravitational vs centrifugal, which hold your ship in weightless orbit. The grip of gravity is now in full control of your descent to earth. Prior to your entry into the earths upper atmosphere, the only braking mechanism your orbiter will have is that caused by the increasing density of gases that envelope earth. Even under the rarified conditions of the exosphere, due to the very large velocity of your ship in this comparative vacuum, your ship experiences drag and begins to slow down and glow.

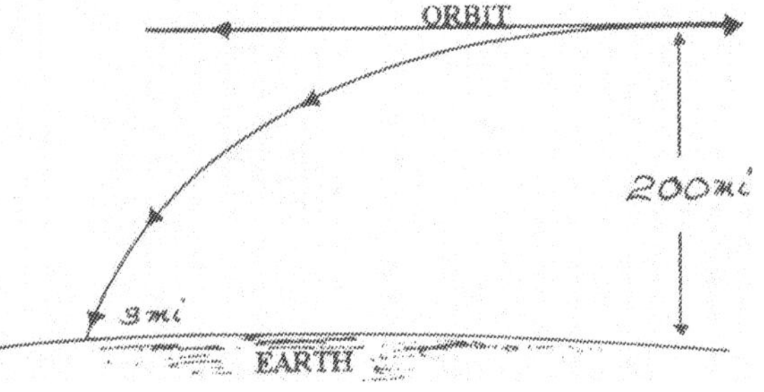

The following is a seven phase hypothetical, but understandable, mathematical description of the conditions that take you and your re-entry vehicle to within three miles of the earth's surface. Each phase is 60 seconds long and opts an analogue of the conditions of space resistance to the motion of your craft.

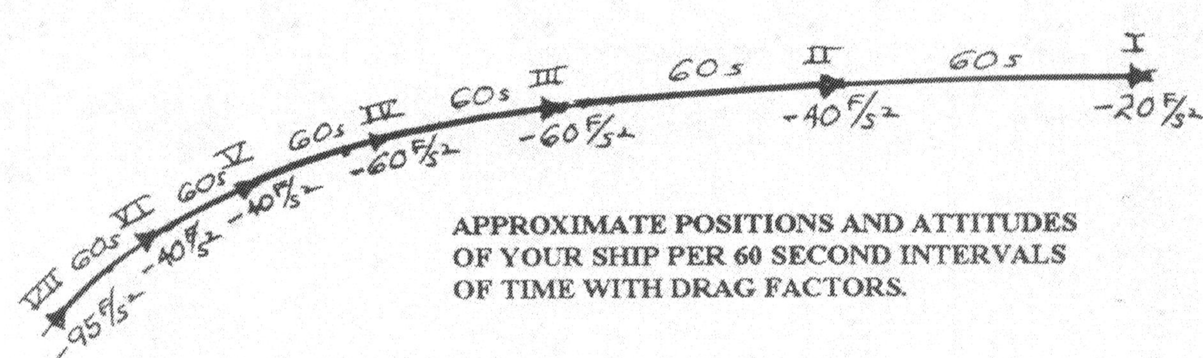

APPROXIMATE POSITIONS AND ATTITUDES OF YOUR SHIP PER 60 SECOND INTERVALS OF TIME WITH DRAG FACTORS.

The conditions of resistance act perpendicularly to the motion of the ship to retard its forward motion. The greater the velocity of the orbiter, the greater the drag effect. Each analogue is represented by a factor of negative acceleration which reduces the velocity of the ship per second. For example, Phase I = -20/ft/s-s. This reduces the forward motion of the ship by subtracting (Cosθ x 20 f/s) from the Vx component and (Sinθ x -20 f/s) form the Vy component per second of travel. This is because the ship is traveling in an arc with a constantly changing angle. For this reason, vectors and angles must be used to insure greater accuracy in the analogue which is being used. Were the ship traveling in a straight line, then 20 f/s would be subtracted from the velocity each second for 60 seconds.

Given the retro-fire information for the first 30 seconds of fire, using the value of the downward acceleration to be 14.43 f/s-s, the orbiter dropped 6,494 feet from its original position 200 miles above earth. The new distance form earth's center is 4198.8 miles and the new gravity is 29 f/s-s.

THE DESCENT – PHASES I AND II

It is not possible to develop the mathematics of all of the phases. You will better understand how the phases work in tandem if you see two consecutive phases developed side by side. In this way you can understand how the information resulting from Phase I is used to complete the Phase II calculations, and then to better understand how Phase II data is used to complete Phase III calculations, etc.

DATA from out of orbit retro-fire

DATA from Phase I Calculations

ΔG = 14.43 f/s-s	Functional gravity	ΔG = 16.45 f/s-s
R = 4199 miles	Distance to earth center	R = 4189 miles
G = 29.04 f/s-s	Gravity at ship	G = 29.18 f/s-s
Vy = 433 f/s	Vertical component	Vy = 1270 f/s
Vx = 17985 f/s	Horizontal component	Vx = 16785 f/s
Sx = 123 miles	Horizontal component	Sx = 321 miles
Sy = 1.23 miles	Vertical component	Sy = 9.5 miles

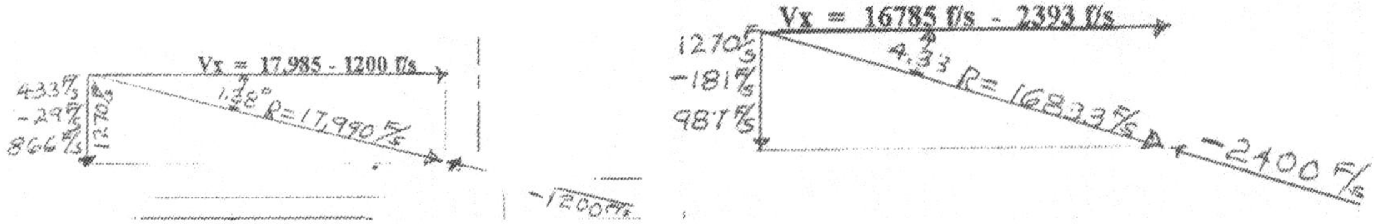

Calculate the resultant velocity for the above and the included angle with horizontal.

R=$\sqrt{(433 f/s)^2 - (17985 f/s)^2} = 17990 f/s$

R =$\sqrt{(1270 f/s)^2 - (17985 f/s)^2} = 16{,}833 f/s$

Cos θ = $\dfrac{17{,}985 f/s}{17{,}990 f/s} \rightarrow \theta = 1.38$ degrees

Cos θ = $\dfrac{16{,}785 f/s}{16{,}833 f/s} \rightarrow \theta = 4.33$ degrees

Calculate the DRAG components for Vx and Vy in the above vector diagram.
xDRAG=(Cos1.38)(-1200f/s)=-1200f/s xDRAG=(Cos4.33)(-2400f/s)=-2393f/s
yDRAG=(Sin1.38)(-1200f/s)=-29f/s yDRAG=(Sin4.33)(-2400f/s)=-181f/s
Calculate the operating gravity factor which incorporates centrifugal force.

ΔG=14.43 f/s-s[This is unchanged] ΔG=29.18 f/s-$\dfrac{(16785 f/s)^2}{(4189 mi)(5280 f/mi)} = 16.45 f/s-s$

Calculate the Vy component (Vx shown in vector drawing).
Vy = (14.43 f/s-s)(60s) = 866 f/s Vy = (16.45 f/s-s)(60s) = 987 f/s

Phase I	**Phase II**
Calculate the total Sx horizontal distance the ship has traveled in both phases.	
Sx=123mi+.5(17985f/s+16785f/s)(60s)=	Sx=321mi+.5(16785f/s+14392f/s)(60s)=
Sx=123mi+198mi=321miles	Sx=321+177mi=498miles
Calculate the total Sy vertical distance the ship has traveled in both phases.	
Sy=(433f/s)(60s)+.5(14.43)(60s)^2… …-(29f/s)(60s)=9.5mi.	Sy=(1270f/s)(60s)+.5(16.45f/s-s)(60s)^2=… …-(181f/s)(60s)=18mi.
Calculate the new distance form the orbital position to the center of earth.	
R = 4199mi-9.5mi=4189miles	R = 4189mi-18mi=4171miles
Calculate the new centrifugal acceleration at this distance from earth center.	
Aorb=$\dfrac{V2}{R} = \dfrac{(16785 r/s)2}{(4189 mi)(5280 f/mi)} 12.73 f/s-s$	Aorb need not be used form this poitn on, because of the larger angle of descent.

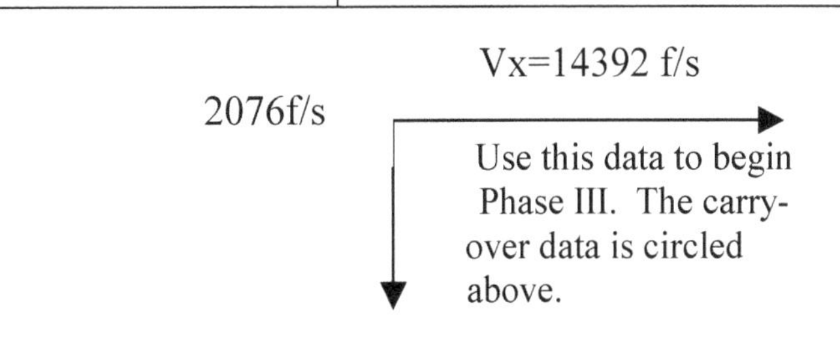

Observe that you have been dropping for only 2.5 minutes and have already descended to 189 miles above earth. Your ship has traveled 498 mi in this short time. As centripetal acceleration will no longer have a significant effect upon the ship, you will now begin to drop very fast.

This is the end of detailed work on phases. However, it is important that you see the results of Phases IV – VII so that you can see how the angle changes as the horizontal components decrease and the vertical components increase. It is also important that you see how the data changes form Phase IV – Phase VII. (The results from Phase III are not provided because you are going to do the calculations for this phase in a later exercise.) The data for these phases are on the next page under the depicted positions of the ship.

Exercise 23

Self Test

You are in an orbiter craft at 300 miles above earth. You are about to retro-fire the ship to begin your descent to earth. The orbiter weighs 3000p, carries 2000p of fuel of thrust capability 40,000p and impulse 350 seconds. Make the following calculations and maintain a data chart of pertinent information along the right side of your paper. (Ref. Page (80))

(1) Calculate the velocity (Vo) of orbit.
(2) Calculate gravity (Gx) at 300 miles above earth.
(3) Calculate the time (T) to burn half of the fuel.
(4) Retro-fire your orbiter burning half of the fuel. Calculate the velocity of gas ejection (Vgas)
(5) Calculate the burn to weight ratio (R).
(6) Calculate the final retro velocity (V) of the ship.
(7) What is the resultant velocity (Vr) of the ship after burn?
(8) What is the weight (iWt) of your ship after burn?
(9) Calculate the CF of the ship after burn.
(10) Calculate the net downward acceleration (A) of the ship at end of burn.

H = 300miles
Wt = 3000p
Wf = 2000p
F = 40,000p
I = 350sec.
Vg =
Gx =
T =
R =
V =
Vr =
IWt =
CF =
A =

Important Data for Phases IV-VII

The calculations for Phases 4-7 are the same as those for Phases 1-3, except that centrifugal force and acceleration need not be taken into consideration after Phases 1 and 2. The re-entry figures below also include the angles, so that you can see the change in orientation of your ship as it descends.

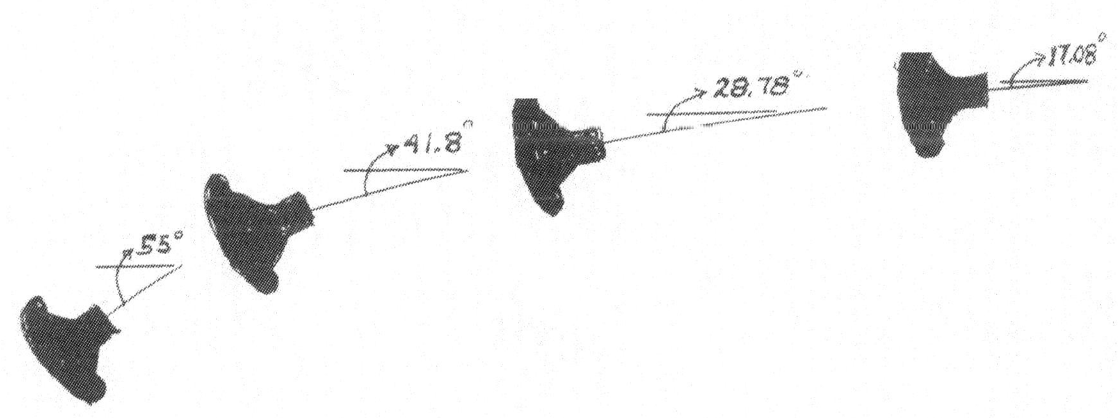

Phase VII Data	Phase VI Data	Phase V Data	Phase IV Data
G = 31.71 f/s-s	G = 31 f/s-s	G = 30.3 f/s-s	G = 29.8 f/s-s
R = 4001.1 mi	R = 4018.3 mi	R = 4064 mi	R = 4107.7 mi
Vy = 1903 f/s	Vy = 1860 f/s	Vy = 1820 f/s	Vy = 1789 f/s
Vx = 399 f/s	Vx = 3496 f/s	Vx = 5285 f/s	Vx = 7388 f/s
Sy = 17.2 mi	Sy = 46.1 mi	Sy = 43.34 mi	Sy = 36 mi
Sx = 22.14 mi	Sx = 49.9 mi	Sx = 72 mi	Sx = 103.5 mi

Looking at the included data, you can see the kinds of changes in radial distance to the center of earth and changes in velocity that you would expect, although a more careful investigation that you will make later on will probably surprise you. You are going to be asked to do. Exercise (24) below, so that you can better understand some of the more subtle details involved here.

Exercise 24

Self Test

Use graph paper to make up a graph of the above information plus the data form Phases 1 and 2. Use the horizontal axis for T and graph G, R, Vy, Vx, Sy, Sx, vs T. Remember that the intervals are 60 seconds. You may place two graphs on the same set of axes. Don't forget to label all axes and the lines representing your graphs. When the graphs are finished, extrapolate (infer) the results of Phase III from your graphs. Keep this information and refer back to it when you do the calculations for Phase III. If you do your graphs accurately, you will be surprised at how nearly your calculated answers agree with the information taken form your graphs, ie., your graphs will tell you whether or not your calculated answers are correct.

The Last Three Miles

Your ship is about to enter the three mile elevation. Drogues have been out and have dropped away and chutes are out. Your horizontal descent has been correct so that you are now falling vertically and your horizontal velocity is now correct to 1535 f/s. You are nearer to the earth's surface than the top of Mt. Everest and will soon reach a constant rate of descent in the earth's atmosphere. Welcome back home and SPLASH DOWN in the Atlantic Ocean.

Your final exercise on this chapter will be to complete the calculations on Phase III of your descent to earth.

Exercise 25

Self Test

Solve Phase III of the re-entry sequence page (81). Make up a table of data for G, R, Vy, Vx, Sy and Sx. Show all work. Discuss the amount of agreement of these values with those form your graphs in Ex. (33).

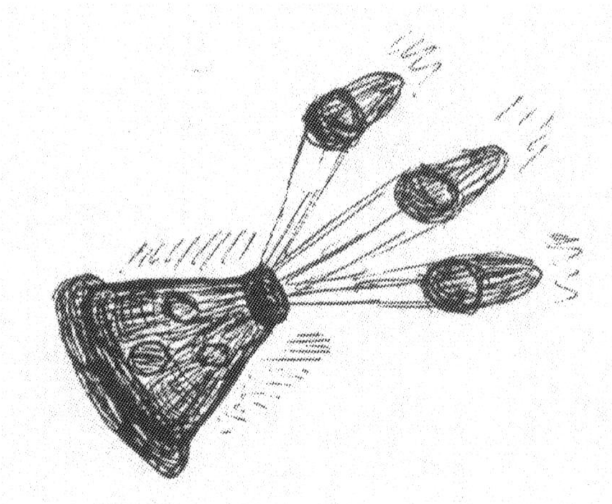

Drogues pop out at between 5 mi. 7 mi. Their purpose is to stabilize the craft and to slow it down in preparation for parachutes. The chutes further slow the entry vehicle for a safe descent and final splash-down.

Chapter VII

The Basic Computer Language

Modern computing, ie., very fast, accurate and involved computerized computation, evolved as a supporting factor to space exploration and navigation. Your hand held calculator will take you through most of the calculation that you will undertake in this program. However, there will soon be some calculations that will be too involved for your calculator. For example, in the next chapter on the Apollo Mission, after the third stage of the Saturn rocket has fired, the ship will drift in space toward the moon for three days. For three days, every second, gravity will be diminished by a very small amount and the ship will have its velocity reduced by a proportionately small amount. If you were to try to use your calculator to make all of these calculations, how many would you have to make?

(3 days)(24 hours/day)(3600 sec/hr) = _________

If you were to try to do that kind of calculation with your calculator, it would take you about five times longer than three days, because you cannot move that fast with your fingers and hands. The problem is to devise a computer language that will make you master of your computer, so that it will do the work you want it to do. Fortunately, there is already such a language. Engineers and scientist use it all of the time. It is called "Basic or Basica."

If your computer already has this language installed, you will find it located under DOS commands. First access DOS, they type Basic or Basica. It will appear on your screen with access commands along the bottom of your screen. Hopefully, you already know about this and know how to use it. If you have the Basic program, but have never used Basic, or if you do not have it on your PC, the following discussion can be easily understood. You should read through this discussion of Basic so as to be familiar with the content of Basic programs. <u>You can skip this chapter on Basic and begin reading about the Apollo 11 Mission into space.</u> For those of you who already know Basic, this will be a good review and a good resource for writing your own programs. For many of you, this will probably be your only opportunity to begin to learn the Basic Computer Language.

<u>Direct Mode Calculations</u>

Basic is a computer language, with its own command vocabulary, which permits you to communicate with your computer in either a <u>direct mode</u> or a <u>program mode</u>. You have already been introduced to the program mode in the section on "achieving orbit." If you have accessed Basic in your PC, input the following commands and observe that pressing 'enter' will give the answers to the following calculator problems:

PRINT 7+5 (ENTER)	PRINT 32-11 (ENTER)	PRINT 4*9 (ENTER)
PRINT 14/3 (ENTER)	PRINT 6^3 (ENTER)	PRINT (36.25) (ENTER)

By now you have discovered that in addition to signed numbers and operations, the basic symbol used for multiplication is (*), for division is (/) and for exponents is (^).

<u>Complex Operations</u>

In doing complex operations, apply the following rules:

1. In programming such operations, an expression can only be written on one line.

2. As in algebraic operations, bear in mind that exponential operations are done first, followed by multiplication and division, then do addition and subtraction.
3. The priority of the above operations can be changed by the use of parentheses. In general, where groups of parentheses are involved, work should be done from the inside paired parentheses outward.

An example of these rules is as follows:
PRINT (3.72^3-33.92)/(9.43-6.59)^3=0.7665509
What answer to the above problem did you get?

Exercise 26

Self Test

Make the following direct mode calculations using Basic.

1. Add 55.55 to 189.22 and subtract 144.651
2. Multiply 359.23 by 67.7.
3. Divide 9.763 by 67.7.
4. Square 38.37
5. Take the cube root of 6,412.
6. Raise 3.77 to the 4.32 power.
7. Use parentheses to write the following: 981 divided by 37 added to 267 divided by 43. Write your answer.

DATA

Data is information from which conclusions can be drawn. In BASIC there are two types of data, ie., NUMERIC and STRING. NUMERIC DATA is composed of integers and real numbers. Numeric data can be stored in the form of SINGLE precision numbers that are accurate to seven places and DOUBLE precision numbers that are accurate to sixteen places. Most of your work will be of the single precision type, as double precision uses up twice as much memory space.

PROGRAMMING

In order to do programming, it will be to your advantage to use ALGORITHMS. An algorithm is a set of instructions that determine the sequence in which individual operations must be performed in order for a computer problem to be solved. If the problem is simple, the algorithm may be in your mind and you type the set of Basic instructions that will solve the problem. You may think of the algorithm as being similar to an algebraic equation which you devise, which uses a set of variables to arrive at a given numerical solution.

VARIABLES

Variables are used in computer programs to solve complex computer problems. Variables are names used to set aside blocks of information that will store data that is to be used in computer programming. Examples of variables are as follows: A, B, C, X, Y, DISTANCE, VELOCITY, ETC.

PRINT and LPRINT STATEMENTS

Already you have used the PRINT statement to output numerical answers. The PRINT statement is the program OUTPUT statement. Typing PRINT will output information on your screen. Typing LPRINT will print information on paper. Without this statement, you will get no printed or readable results.

Your First Program

Your first program will use numbers to designate program statements. You must use sequential numbers, ie., 10, 20, 30, etc. to designate computer statements. Units of ten are used so that there will be room between these numbers to add additional numbers. The purpose of this program will be to designate variable names of three numbers, to add these numbers and print out the result on your screen.

```
10 A = 4.12
20 B = 11.7
30 C = 18.55
40 T = A + B + C
50 LPRINT A, B, C, T
60 STOP
```

PRINTOUT: Point the arrow at RUN on your screen or click the second function key.

```
4.12          11.7          18.55          34.37
```

The program printed to the right of this paragraph has some additional features you may want to try. By using the LPRINT statement followed by paired quotation marks, you can have any given statement included in your program print-out. Three such statements are inserted before (10). You can insert information into your program by typing the statement below the program. It will be included when you activate F(1) to LIST the entire program. Statement (45) which lists the variables to be printed is included in the same form. The printout of the new program is printed below the listed program. If you type LLIST at the bottom of your program, it will immediately be printed by your printer.

```
4 LPRINT "This is a program to add"
6 LPRINT "three variables and to"
8 LPRINT "print the result"
10 A = 4.12
20 B - 11.7
30 C = 18.55
40 T = A + B + C
45 LPRINT "A      B       C       T"
50 LPRINT A,B,C,T
60 STOP

This is a program to add three
variables and to print the result
A     B        C          T
4.12  11.7     18.55      34.37
```

NON PROGRAM Statements

There are items when you want to explain the use of some of your program statements for purposes of clarification, but you do not want the statement to become part of your written program. The use of the word REM, or the apostrophe (‘), will do this for you. The following is an example of this:

```
10 REM This statement will not be included in the printout
20 ‘And this is not part of the program either
```

ASSIGNMENT INSTRUCTIONS

Assignment instructions are used to assign values to variables: A = 4.12, B = 11.7; C = D; VEL = 25; AGE = 15; all are examples of assignment instructions. In computer language, a becomes 4.12, and VEL becomes 25.

Exercise 27

Self Test

Devise a program in which you use three variables with assigned values. Multiply two of them and add the third variable to their sum. Use the same kid of format as in the above program and printout. Include at least one quotation statement and one that will not be in the printout.

More Information on Variables

Variables are very important to programming. In addition to those we have discussed, there are three special variables used with numbers. These are (%), (!), and (#). The following are examples of how numeric variables are designated: A%; B!; D#. These symbols can be used with a combination of letters and numbers. The symbol (%) stands for integer numbers. The symbols (!) and (#) are used with decimal numerals. (!) stands for SINGLE PRECISION numbers and provides an accuracy to seven places. (#) is used to represent DOUBLE PRECISION numbers accurate to sixteen places. The following is a test of the validity of these statements:

```
5 REM This is an example of how these symbols are used
10 A% = 23.52
20 B! = 29.765437825
30 C# = 29.765437825
40 PRINT A%, B!, C#
50 END
```

When you ran this program, what did you discover? Integer really does mean whole numbers. (!) will print out only seven places. (#) will accurately store all numbers that you entered as data. All of these tools are powerful in making numbers do what you want them to do, but if you use wrong symbols, the computer will provide you with wrong numbers.

INPUT Statements

In all of the program s that you have done, you had to pre-type the information at the beginning of the program. Another way to put data into a program is to use the INPUT statement and type in the information when your program progresses to where it is needed.

FORMATTING

PRINT statements which have two or more parameters which are separated by (;) will be printed on the same line and separated by spaces. EXAMPLE:

```
10 PRINT "Armstrong "; "Aldrin  "; "Carter"
20 END
```

Check the above to see if the premise is true.

Consecutive PRINT statements that end with a semi colon will be printed on the same line until space has run out and then will be continued on the next line.

PRINT statements in which the (,) is used will set up columns in your printing. The columns start at 1, 15, 29, 43, 57…to 80 places. Therefore ( ) = 1; (,) = 15; (,,) = 29; (,,,) = 43, etc., to 80 spaces.

EXAMPLE:

```
10 RUN THIS PROGRAM ON YOUR COMPUTER TO CHECK VALIDITY
20 PRINT "Force"
30 PRINT ,"Fuel"
40 PRINT ,,"Impulse"
50 PRINT ,,,"Thrust"
60 PRINT ,,,,"Burn"
70 END
```

The PRINT TAB (); function is the last method of controlling the placement of data on a page. As you are aware, there are 80 spaces to a line. You can use the TAB function to place data or statements on a line where you wish them to be. For example, TAB (18) will place the first letter of the word at that point on the line. EXAMPLE:

```
10 REM Be sure to run this program so that you will be able to use
20 REM the TAB statement to place output data.
30 PRINT TAB (12); "SOLID";
40 PRINT TAB (21); "AND";
50 PRINT TAB (29);"LIQUID";
60 END
```

Exercise 28

Self Test

Given the words Atlas, Titan and Vanguard, devise a program for each condition below. Use the comma and/or the semi colon to place these words in a printout. Submit the program with results.

PROGRAM I write a print statement on three different program lines in which these words will be printed on a single line.

The following are three short programs that include GO TO statements:

PROGRAM II Write a print statement in which your program will be on a single line, all words equally spaced.

PROGRAM III Write print statements in which your program will be output on three lines, but set up in equally spaced columns.

PROGRAM IV Write a program, using TABS, in which you place the words on the same line, each word placed at twice the spacing of the previous word.

GO TO STATEMENT

This will be your introduction to LOOPS—a very powerful tool in writing computer programs. The GO TO statement lets you direct the computer to any statement in your program. The statement by itself, directed to a given point in your program, will cause the program to go into a loop, always going back to the same place. In order to be effective, it must be written as a conditional statement, but more about that later. It can be used to direct the program to a print statement or to the end of the program to a STOP statement. The following are three short programs that include GO TO statements:

10 A = 5	10 A = 5	10 A = 5
20 A = A + 1	20 A = A + 1	20 A = A + 1
30 PRINT A	30 PRINT A	30 PRINT A
40 GO TO 20	40 GO TO 50	40 If A > 11 THEN 60
50 STOP	50 STOP	50 GOTO TO 20
		60 STOP

Observe that in the first program, statement 40 directs the program back to 20. This program is in a continuous loop and will go on adding 1 to each previous A.

In the second program, statement 40 directs the program to statement 50 and the program stops after it prints (6).

In the third program, statement 40 is conditional, so the program will pass to statement 50 and add 1 to the previous A until its total is 12, which is the first number greater than 11. At that point, the program is direct to statement 60 and the program stops. You do not want any program to run a continuous loop, as you must shut down the machine to stop the program. In the second program, you do not need the GO TO statement, because the program would cycle only once and stop without the STOP statement. You can see that there is great advantage to being able to use the conditional statement, because that gives you full control over the program.

USING THE IF STATEMENT

Almost as big as "life" itself is the concept of relative size and worth. It is a natural thing to be "taller," "bigger," "faster," than another person. This is the essence of competition and growing up. The IF statement is powerful because it permits you to compare all types of things and conditions. "Rockets are faster than airplanes," can be expressed as 'rocket velocity>than airplane velocity. Conversely, airplane velocity<rocket velocity. Other conditions related to velocity of travel could be expressed with inequality statements such as (<=) less than or equal to;(>=) greater than or equal to; or (<>) not equal to. These are LOGICAL EXPRESSIONS and have either a true or false condition, or 'yes or no' condition, or in Boolean algebra (1) or (0) as finite values.

The IF, THEN statement is a conditional expression in which the condition is the inequality the statement is action to be carried out by the program. Examples are as follows:

Condition	Statement
IF your speed > 75	THEN PRINT "Slow down"
IF your speed < 30	THEN PRINT "Speed up"
IF you speed <> 45	THEN PRINT "Continue on"

The following program, with printout, illustrates the use of the IF THEN statement and will further acquaint you with printing techniques.

```
10 'This is an example of a GO TO statement used in conjunction
20 'with an IF THEN statement.
30 LPRINT "V=Velocity in miles per hour"
40 V=0
50 V=V+10
60 'Statement 50 is an example of an adder
70 IF V=30 THEN LPRINT, "V=30, SPEED UP"
80 IF V+&) THEN LPRINT,,"V=70, SLOW DOWN"
85 'Note the use of commas in the print statements
90 IF V=90 THEN GO TO 140
100 'Statement (90) takes the program out of the LOOP
110 GO TO 50
120 'Statement 110 creates the LOOP by sending the program back to
130 'Statement 50 to initiate the program again
140 LPRINT V
150 STOP
```

```
V = 30, SPEED UP
                    V = 70, SLOW DOWN
90
```

MORE INFORMATION ON LOOPS

I cannot over-emphasize the importance of LOOPS. The FOR...NEXT statement is the prompt for a LOOP. The statement, which used the integer symbol (%) is as follows:

FOR S% = 2 to 10 STEP 2

This statement will process the numbers 2, 4, 6...10. The following program with printout is an example of the application of the above statement to Basic programming:

```
10 FOR S% = 2 to 10 STEP 2 'This is FOR and INCREMENT
20 PRINT "S% =  ";S%
30 NEXT S%   'The NEXT statement initiates the LOOP
40 END
```

PRINTOUT; S%=2, S%=4, S%=6, S%=8, S%=10

As you can see, the program will move through the LOOP five times, incrementing the first 2 and then 4 by 2 until the last number 10. It will print each step of the loop prompted by NEXT and the END the program.

Exercise 29

Self Test

Write a FOR NEXT program that will process umbers from 3 to 21, incrementing the numbers by 3, but will end the program prior to reaching the number 18. Include in the program an LPRINT statement that will list your output. Use an LLIST statement to print the program.

Using a FOR NEXT LOOP to INPUT information

The INPUT statement can be used in any of your programs. It is the one statement that permits you to enter information while the program is processing. The following program progresses through the loop four times. Each time it passes through the loop it requests your INPUT information before it can continue on through the loop. In this program, the loop numbers act as counters and establish the number of times the loop is processed.

```
10 SUM=0;PROD=1 'Defines the lowest value for SUM and PROD
20 FOR T%=1 to 4 'Defines the number of loops
30 INPUT "ANY NUMBER:  ",N
40 SUM = SUM + N
50 PROD = PROD = N
60 PRINT "N =  ";N
70 NEXT T% 'Transfers processing back to the FOR T%  statement
80 PRINT "SUM =  "; SUM; "PROD =  ";PROD
90 END
```

```
PRINTOUT
N = 25
N = 35
N = 45
N = 55
SUM = 100   PROD = 2165625
```

The Use of FORMULAE IN LOOPS

Think of this as the beginning of FUNCTION applications in programming. In the next program, a loop is used to process the 'free fall' formula five times. The real application of this is in the laboratory where you are making many calculations of the same sequence. The more complex the equation, the greater the resolving power of the loop. The formula to be used in the program is:

```
S = (1/2)(G)(T)
REM (G) is 32 ft/sec-sec and (T), the loop value, is in seconds.
10 G! = 32
20 FOR T! = 1 TO 5
30 S! = .5 * G! = T! – 2
40 PRINT, "T!"; T!, "S!"; S!
50 NEXT T:
60 END
```

PRINTOUT

T! 1	S! 16
T! 2	S! 64
T! 3	S! 144
T! 4	S! 256
T! 5	S! 400

READ AND DATA STATEMENTS

READ and DATA statements offer a very important way to enter data into your computer. This data can be used only once (this is one time input data). The words READ and DATA always go together in a program, although they do not necessarily follow each other. You may have one or more READ statements, but you must have at least one DATA statement. The READ statement is as follows: (Variables and values are arranged one above the other so as to show a one on one correspondence.

READ VAR1, VAR2, VAR3.....VARX. The DATA statement is as follows:

DATA VAL1, VAL2, VAL3....VALX.

Example of a READ statement: READ dist, vel, accel, g

Example of a DATA statement: DATA 550, 75, 7.52, 32

```
EXAMPLE:  10 DATA 1, 2, 3, 4, 5 'This statement declares values
20 READ A, B, C, D, E 'Reads the values from data
30 PRINT A, B, C, D, E
40 END
PRINTOUT
          1    2    3    4    5
```

Exercise 30

Self Test

Write a program in which you use a READ and DATA statement that includes four variables and four values, and a formula statement that uses the variables and values. Include a PRINT statement that will list the variables, values and a derived result.

MATHEMATICAL FUNCTIONS (Trigonometric)

From the recent work you did on re-entry and landing procedures, you know the importance of using trigonometric functions. Using these functions as part of a computer program is more difficult than using them with a scientific calculator.

TRIGONOMETRIC COMPUTER FUNCTIONS

There are four trig functions which you will learn to use. These are Sine(SIN), Cosine(COS), Tangent(TAN) and arctangent(ATN) functions. The most difficult part of their computer application is that the angle must be converted to radian measure. Radian measure was discussed in the physics section, so you already have previous information on this subject.

RADIAN MEASURE

The RADIAN is defined as that angle subtended by an arc equal in length to the radius of its circle. That angle is 57.3 degrees. You can always calculate this angle as follows:

1 RADIAN = (180 deg)/(3.1416) = 57.3 degrees

(Of course, there are 2 pi radians in 360 degrees.)

The following program will test the above hypothesis by establishing a loop to define angles, converting the angles in degrees to radians and taking the sine function of the angle, after which it will define and print all values. Note another method for printing data:

```
10 FOR A = 30 to 360 STEP 30 'Defines the angle between 30 and 360
20 R! = 3.1416 * A/180 'Converts the angle to radian measure
30 S! = SIN(R!) 'Provides the sign function of the angle
40 LPRINT "A = "A, "R! =  "R, "S! = "S!
50 NEXT
60 STOP
```

PRINTOUT

A = 30	R! = .5236	S! = .5000011
A = 60	R! = 1.0472	S! = .8660266
A = 90	R! = 1.5708	S! = 1
A = 120	R! = 2.0944	S! = .8660231
A = 150	R! = 2.618	S! = .4999947
A = 180	R! = 3.1416	S! = -7.239981E-06
A = 210	R! = 3.6652	S! = -.5000073
A = 240	R! = 4.1888	S! = -8660302
A = 270	R! = 4.7124	S! = -1
A = 300	R! = 5.236	S! = -.8660192
A = 330	R! = 5.7596	S! = -.4999887
A = 360	R! = 6.2832	S! = 1.447996E-05

Before you go on to LOGARITHMIC FUNCTIONS, check the sine functions of the angles in the last program with your calculator. Are the results correct? To how many places should they be correct?

LOGARITHMIC FUNCTIONS

This will be the last instructional part of the program. Logarithmic functions are to the base (e) where e = 2.718282. These are called natural logarithms. For any positive number (y), the value of the LOG(Y) is its natural logarithm, or log to the base (e). This is the function that you obtained on your scientific calculator when you took the In, of the weight to burn ratio, as part of the final determination of the velocity of the rocket. The computer generated inverse function of LOG is EXP(x). An example program is as follows:

```
10 DATA .6, 1.3, 2.86, .067, 3.942
20 READ A!, B!, C!, D!, E!
30 F! = LOG(A!)
40 G! = LOG(B!)
50 H! = LOG(C!)
60 I! = LOG(D!)
70 J! = LOG(E!)
```

```
80 PRINT A!, B!, C!, D!, E!
90 PRINT F!, G!, H!, I!, J!
100 STOP
PRINTOUT
```

.6	1.3	2.86	.067	3.942
-.5108257	.2623642	1.050822	-2.703063	1.371688

Exercise 31

Self Test

Use the computer's "Basic Direct Mode" function to provide answers for the following problems (Use your calculator to check your answers.)

1. Convert 40 degrees to radians.
2. Convert 80 degrees to radians.
3. Convert 3.92 radians to degrees.
4. SIN 40 degrees = ___________
5. COS 140 degrees = _________
6. TAN 80 degrees = __________
7. Find the ATN of .813 = _________
8. Find the LOG of 5.31 + ___________
9. calculate the anti log (EXP) function of 1.73 ___________
10. Calculate the LOG of 5.641 = ______________
11. What do you think is the relationship between the functions in (9) and (10).

THE FINAL BASIC PROGRAM

This is the final program of this chapter. Its purpose is to demonstrate the use of SIN and LOG functions on a rocket problem, but also to show the importance of the angle of a rocket to its perpendicular with earth as this relates to the gravitational drag on the velocity of the rocket. First the program will determine the optimum velocity of the rocket at burnout given the exhaust velocity VEX = 8030 ft/sec; ship's total weight WTS = 1,411,000 pounds; first stage after burn weight WBN = 351,000 p. The program will then explore the change in this velocity VROC, as the angle with the perpendicular to earth changes form 0 degrees to 90 degrees, as the rocket moves from earth position to orbital position. VRES is the calculated final resultant velocity of the rocket.

THE BASIC PROGRAM

```
10 LPRINT "A = Angle"
20 LPRINT "VEX = Exhaust velocity"
30 LPRINT "WTS = Weight of rocket before burn"
40 LPRINT "WAB = Weight of rocket after burn"
50 LPRINT "IWB = Fuel burn per 10 degrees of flight"
```

```
60 LPRINT "T = Total burn time"
70 LPRINT "G = Gravity (G changes very little during this flight period)"
80 LPRINT "IT = Incremental time per 10 degrees of flight"
90 LPRINT "AGT = Gravitational drag per 10 degrees of flight"
110 LPRINT
120 LPRINT "(A) Angle   (VR) Resultant Velocity  (AGT) Drag"
130 LPRINT
140 DATA 8030, 1411000, 133, 351000, 32
150 READ VEX!, WTS!, T, WAB!, G
160 WTS! = 1411000!
170 IT! = 133/9 'This is the calculation of incremental time
180 N = 0
190 IWB = (WTS! – 351000!)/9 'This is the calc. Of fuel burned per 10 degrees
200 FOR A = 10 to 90 STEP 10 'This begins the loop
210 N = N + 1
220 R! = WTS!/(WTS! – N * IWB!) 'This is the weight ratio calculation
230 AGT! = N * IT! * 32 * COS(A/57.3) 'This is the drag calculation
240 VR! = 0030 * LOG(R!) – AGT! 'This is the calc. Of VR per 10 degrees
250 LPRINT "A =  "A, "VR! =  "VR!,,"AGT! =  "AGT!
260 NEXT A
270 END
A = Angle
VEX = Exhaust velocity
WTS  = Weight of rocket before burn
WAB = Weight of rocket after burn
IWB = Fuel burn per 10 degrees of flight
T = Total burn time
G = Gravity (G changes very little during this flight period)
IT = Incremental time per 10 degrees of flight
AGT = Gravitational drag per 10 degrees of flight
```

PROGRAM PRINTOUT

(A) Angle	(VR) Resultant Velocity	(AGT) Drag
A = 10	VR! = 234.2017	AGT! = 465.7057
A = 20	VR! = 577.9491	AGT! = 888.7487
A = 30	VR! = 1085.886	AGT! = 1228.629
A = 40	VR! = 1813.449	AGT! = 1449.078
A = 50	VR! = 2817.68	AGT! = 1519.952
A = 60	VR! = 4160.404	AGT! = 1418.856
A = 70	VR! = 5916.184	AGT! = 1132.443
A = 80	VR! = 8191.146	AGT! = 657.3134
A = 90	VR! = 11171.39	AGT! = .4924554

This Basic Program and data printout has much important information related to programming and to rocket theory. Before leaving this program it is important that you take a careful look at it and its results. Within the printed data is important information about rocket acceleration from the pad, and the functional change in gravity due to the angle of rocket flight. The following exercise will deal with these relationships:

Exercise 32

Self Test

1 and 2. Graph A vs VR and A vs AGT on the same axis. Use the x axis for A values.
3. Discuss what your graphs tell you about the interrelationship between VR and AGT as the angle increases.
4. Looking at the first three (VR) values, what is the approximate ratio of increase between (a) the first to the second, and (b) the second to the third sets of data? (c) Are these increases about the same?
5. Looking at the last three (VR) values, what was the increase in velocity from (a) the seventh to the eight velocity in ft/sec, and (b), from the eighth to the ninth value in ft/sec? (c) What, if anything, has this to do with the angle of flight.
6. Looking at the (AGT) data, (a), what do you observe happened to the drag characteristic of the rocket? (b) At what angle was there an observable change in this characteristic? (c) The functional change in what physical characteristic, brought about his change?
7. Looking at the program, write the formula for the "fuel burn per 10" calculation.
8. Use the program to write the formula for the Ratio calculation.
9. Use the program to write the formula for the AGT calculation.
10. Write the formula for the VR calculation using the basic program.

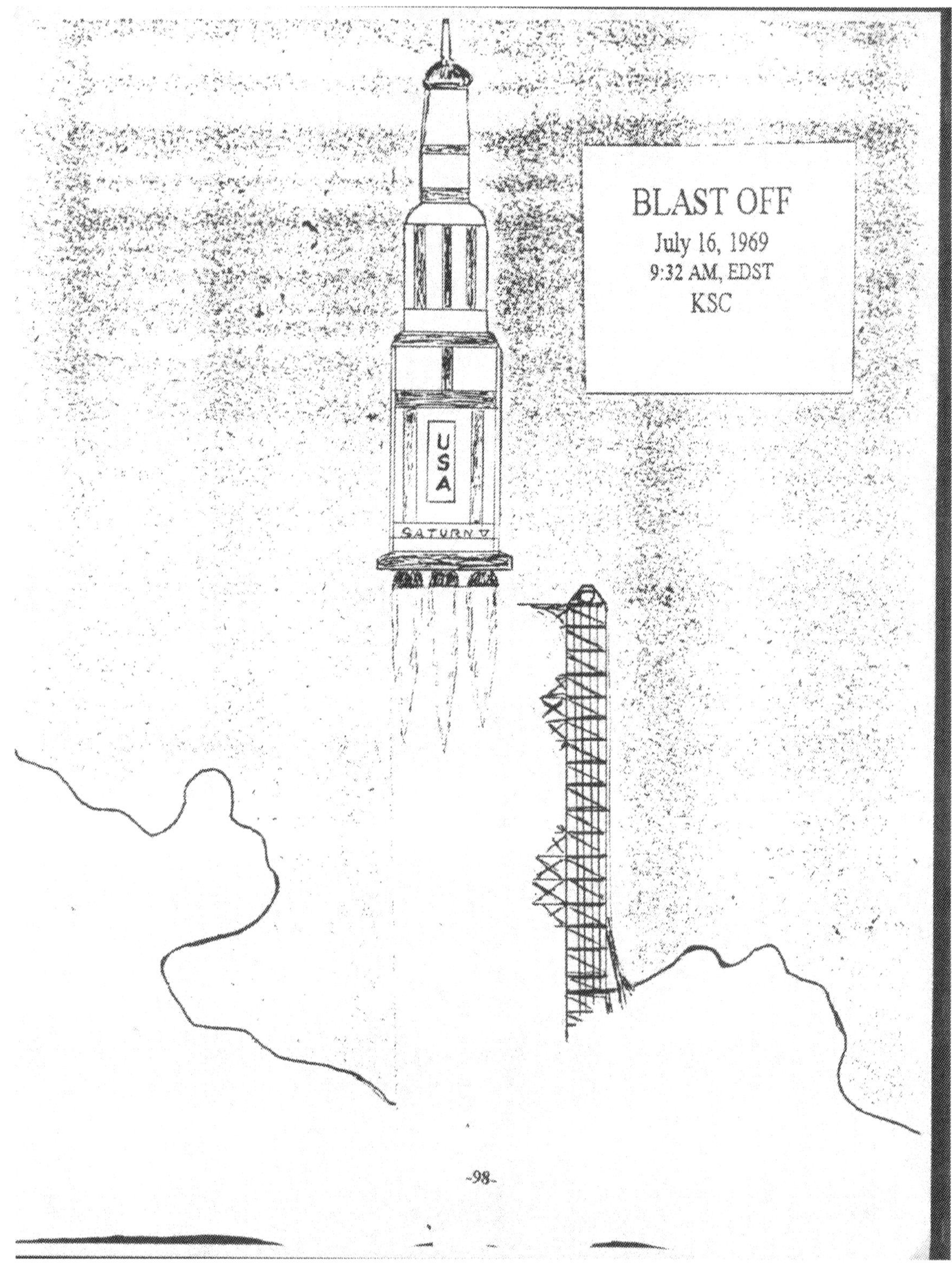
BLAST OFF
July 16, 1969
9:32 AM, EDST
KSC
USA
SATURN V
-98-

PART III – The Apollo 11 Mission

Chapter VIII

Earth – Gravity – Maneuvers

The International Race

As late as 1959 and into the early 60s, the space program of the USSR was still several years ahead of the United States. In January 1959, the Soviet Union launched Luna I, which passed by the moon and went into orbit around the sun. In September, that same year, Luna II was launched and impacted the moon. In October, Luna III orbited the moon at 40,000 miles and took the first pictures ever taken of the back side of the moon. For that reason, most of the features of that side of the moon were given Russian names. In August of 1960, the Russians placed two dogs in orbit and safely recovered them. These were the first living creatures to be placed in orbit. In April 1961, Yuri Gagarin became the first human space traveler. Vostok I circled the earth at a velocity of 17,400 mph. The single orbit reached a maximum height of 203 miles and lasted 1 hour and 48 minutes.

The US finally entered the space race in May 1961, when Cmdr. Alan Shepard, Jr. was rocketed downrange reaching a top speed of 5,100 mph. His flight time was 15 minutes and he reached an altitude of 115.7 miles. That was only 40 years ago. You can see that the USSR was far ahead of this country at that time. The space program of the US took a quantum leap when, on Dec. 24, 1968, Apollo 8, with Col. Frank Borman, Capt. James Lovell, Jr. and Maj. Wm. Anders, circled the moon ten times and re-entered the earth's atmosphere at 24, 695 mph. It was only seven months later that Apollo 11, manned by Lt. Col. Michael Collins, Neil Armstrong and Col. Edwin Aldrin, landed on the surface of the moon. Their flight time was 195 hr., 18 minutes. The third and final moon landing was made by Maj. Stuart Roosa, Capt. Alan Shepard and Comdr. Edgar Mitchell in Apollo 14. That flight lasted form Jan. 31st to Feb. 9th. With the exception of Apollo 13, those were the first and last times that any human being ever left the gravitational field of this world or ever set foot upon an extraterrestrial body. Even 40 years later, it is doubtful that we could achieve the same objectives, with the same precision, and without the loss of a single life.

It is important that you read about the people and events that were Apollo. But, don't lose sight of the fact that tens of thousands of scientists, engineers and technicians, who knew their jobs very well, who designed the rockets, engines and fuel, who networked the electronic systems to space navigation, provided the coordinated structure that made Apollo successful. All of these people were still working at their jobs after the final moon landing. They continued to dream about an international, permanently orbiting space station, about building a station on the moon and about a manned trip to Mars. Much of their energy went into the successful development of the space shuttle. This was the machine that could make all of their dreams come true. But money became short and politics prevailed. Somehow, all of the dreams ended. When the Challenger blew up in January of 1986, the entire space program shut down for several years.

In 1492, when Columbus set out to find a shortcut to Asia, his greatest enemy was the 'unknown" and the belief among many of his working crew, that the world was not round but was flat. He and other intelligent men of his time knew this to be a false belief. If a storm wrecked his ships, he was still on the same planet and could possibly survive to float to some island where he could live out his life. There was little chance that there would be any ship that would look for him.

In 1969, when the Apollo crew lifted away form Cape Canaveral at 9:32 AM, EDST, on July 16th, they could clearly see where they were going. Even so, the dangers were much greater for them than for Columbus. There could be no survival on the surface of the moon if they crashed. Worse than that, if their trajectory from earth was not correct, they would miss the moon and continue on into space to orbit the sun forever. These three brave men accepted the danger and went on to accomplish their mission in space.

The Saturn Booster

Apollo was an absolute giant among rockets. The primary Saturn V rocket was composed of three stages. The rocket system became Apollo when the lunar component, 58 feet long, weighing 90,000 pounds, was added to the top of the third stage. The entire length of the ship was 363 feet. Apollo 11, as it stood at the gantry ready for man's first mission to the moon, was longer than a football field, and again one third longer than any other rocket system ever assembled. In all, its weight was greater than 6.84 million pounds and its thrust capability was 7.5 million pounds. The men, who occupied the Command Module at the top of the ship, were so dwarfed by its size they looked like nesting insects.

But you cannot leave until Mission Control completes the count-down. Upon completion of the countdown, you will first be fired into a 118 mile orbital trajectory, and then on into outer space where you will escape earth's gravity. However, before you can leave, you must first complete the calculations that will make this trip possible, because all aspects of this journey had to be pre-calculated, computerized and checked out for validity. The count down has been underway for several hours, with many important details yet to be checked. In order to successfully complete these calculations, you must know the following additional data:

(1) The first stage of the rocket contains 4.5 million p of (lox) and kerosene.
(2) The impulse (I) of this fuel is 320 seconds in space.
(3) The empty tank of the first stage weighs 226 tons.
(4) The average first stage angle is 50 degrees with the vertical.

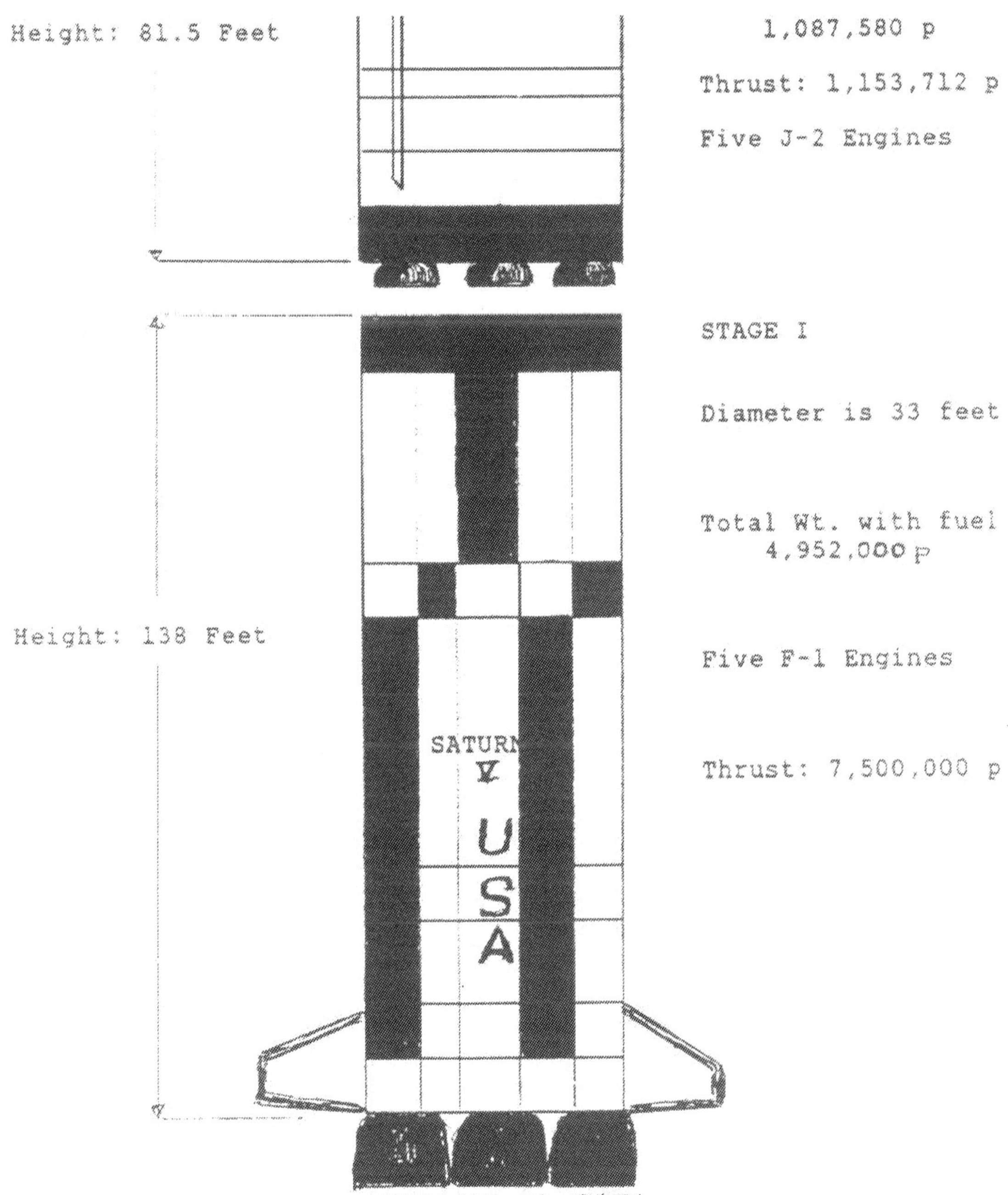

APOLLO 11 DATA

SATURN V, 363 FEET LONG, STANDS AS ONE OF AMERICA'S GREAT ACHIEVEMENTS IN THE FIELD OF ROCKET ENGINEERING. THE DEVELOPMENT OF THIS ROCKET MADE POSSIBLE THE EXPLORATION OF THE MOON. YET TODAY, 1999 IT COULD BE USED FOR THE FURTHER EXPLORATION OF SPACE.

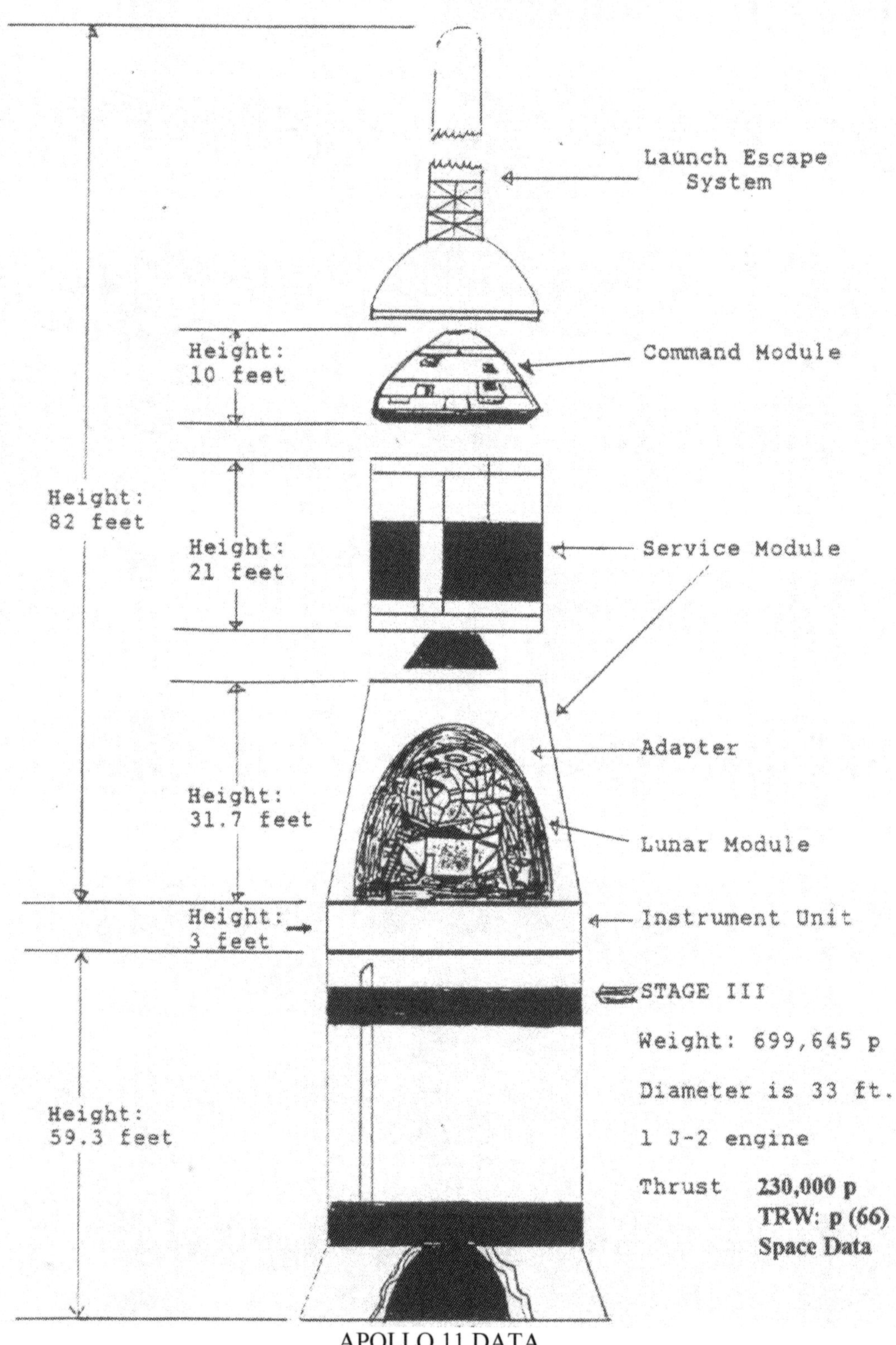

APOLLO 11 DATA

This illustration of the Saturn V Rocket is excellent for cross-sectional detail. The designated data is sufficiently accurate to get you to the moon and back. CM, SM & LM weight is 100,000 pounds.

APOLLO-First Stage Calculations

(1) Calculate the time (t) to burnout using the impulse (I) equation.

$$T=\frac{(Wf)(I)}{(F)}=\frac{(4{,}500{,}000p)(320s)}{7{,}570{,}000p}=190.22s$$

(2) Calculate the velocity of gas ejection.
V=

$$\frac{(Thrust)(T)(g)}{Wt\_of\_fuel}=\frac{(7{,}570{,}000p)(32f/s-s)(190.22s)}{4{,}500{,}000p}$$

(3) Calculate the ratio (R) of the ship's weight before burn to its weight after burn, then calculate the natural log (In) of (R).
R=

$$\frac{(Wt)before\_burn}{(Wt)after\_burn}=\frac{6{,}840{,}000p}{2{,}340{,}000p}=2.93\rightarrow In(R)=1.073$$

(4) Calculate the final velocity of the ship at "after burn".
V=((Vg)(In(R)))-(g)(T)(Cos0)=
V=(10240 f/s)(1.073)-(32 f/s-s)(190s)(Cos50°)=
V=10984 f/s-3908 f/s=7076 f/s

(5) Calculate the ships elevation and distance from its point of origin using <θ=50°. Make a triangular drawing, using correct velocity values.
Vy=V(Sin 40°)=(7076 f/s)(.643)=4550 f/s
Vx=V(Cos 40°)=(7076 f/s)(.766)=5420 f/s
H=.5(Vy)(T)=(2275 f/s)(190s)/(5280 f/mi)=81.87 mi
Sx=.5(Vx)(T)=2710 f/s)(190s)/(5280 f/mi)=97.52 mi

Now that you have completed this information, it will be entered into NASA's computers for the ship's firing and navigational control. MAKE THE FINAL COUNTDOWN TO BLAST-OFF: 6---IGNITION—5—5—3—2—1 "BLAST OFF".

Write the correct data for each item.

Impulse 320s

Fire time 190.22s

Velocity of gas ejection
10,240 f/s

First stage final velocity (ft/sec)
7076 ft/sec

Final vel. (mph)
4825 mi/hr

Altitude at engine cut off
81.87 mi.

Linear distance at engine cut off
97.52 mi.

Pre-fire W of ship
6,840,000 p

Vy component
4550 ft/sec

Vx component
5420 ft/sec

Empty W of first stage tank
452,775 p

Weight of ship at after tk. Drop
1,887,225 p

NASA: (July 16th, 7 min. and 41 sec. Downrange, 530 mi, altitude 95 miles.)

APOLLO-Second Stage Calculation

The second stage of a Saturn V missile uses liquid oxygen and hydrogen, O_2H_2. This high energy fuel has an impulse of 450s in a vacuum, but only 340s in the atmosphere at sea level. The mesosphere begins at 50 miles above seal level and this elevation may be considered a vacuum. The 2nd stage of Saturn V uses five engines, the same as its first stage.

Saturn V, second stage, went into a temporary parking orbit at the end of its final burn, but with a brief third stage maneuver, after which the engine shut down.

Second Stage Calculations are as follows:

$$T=\frac{(W_f)(I)}{F}=\frac{(978{,}000p)(450s)}{1{,}153{,}712p}=381.5\sec$$

Vg=

$$\frac{(F)(g)(T)}{Wf}=\frac{(1{,}153{,}712p(32f/s-s)(381.5s)}{978{,}000}=14{,}401f/s$$

$$R=\frac{(W)total}{(W)burn}=\frac{1{,}887{,}225p}{909225p}=2.08\rightarrow In(2.08)=.73$$

Vii=(Vg)(InR)-(gx)(T)(Cos θ)
Vii=(14401f/s)(.73)-(30.73 f/s-s)(381.5s)(Cos 85°)
Vii=10513 f/s-1022 f/s=9491 f/s

<u>Data –2nd stage</u>

F=<u>1,153,712 p</u>
Ships angle = <u>85°</u>
Wfuel=<u>978,000p</u>
Tburn=<u>381.5s</u>
Vgas=<u>14,401 f/s</u>
Wt (2nd stage)
<u>1,087,580 p</u>
Wt (total)=<u>1,887,225p</u>
<u>Vii=9491 f/s</u> or <u>6471 mi/hr</u>
H(2nd burn)=<u>29.9 mi</u>
S (2nd burn) = <u>341.6 mi</u>
H final altitude <u>111.8 mi</u>
S final distance <u>830.72 miles</u>
G(x) = <u>30.73 f/s-s</u>
Wship after tk. Drop <u>799.645 p</u>
W (2nd stage)Tk 109,580p

Calculate the vertical and horizontal distance gained by the ship in its second burn.
(Vy)calc. For 2nd burn☹Vy=9491 f/s)(Sin 5°)=827.2 f/s
(Vx)calc. For 2nd burn: (Vx=9491 f/s)(Cos 5°)=9455 f/s

(H) calc. 2nd burn: H = (414 f/s)(381.5s)/(5,280 f/mi)=29.9 miles
(Sx) calc. 2nd burn: Sx=(4728 f/s)(381.5s)/(5,280 f/mi)=341.6 miles

Calculate the total horizontal distance the ship has traveled at end of 2nd burn.
Sx = 97.52 mi + 341.6 mi + 391.6 mi = 830.72 miles

NASA: (July 16th, II min and 42 sec., altitude 100 mi, downrange 883 mi.)

Calculate the distance of the ship above earth-2nd burn.
H'=29.9 mi + 81.87 mi = 111.8 miles

Calculate the final velocity of the ship after second burn (use the velocity of earth rotation in this calculation).
ΔV"=7076 f/s + 9491 f/s + 1535 f/s = 18,102 f/s = 12,342 mi/hr

Your ship will achieve orbit at 118 miles above earth. Calculate the 'park velocity.

$$\text{Vpark}=\sqrt{(g)(\text{Re})/Rx}=\sqrt{\frac{(32f/s-s)(4000mi\_x\_5{,}280f/mi)^2}{(4118mi\_x\_5280f/mi)}}=25{,}621f/s$$ or 17,468 mi/hr.

At final second stage burnout, the ship's velocity is lower than that required to remain in parking orbit. How much additional velocity will the ship need?

V = 25,621 f/s – 18,102 f/s = 7519 f/s

The above velocity will require an additional expenditure of 326,699 p of fuel form the third stage tank. The burn-time is 639.2 seconds. Later on, you will learn how to calculate the amount of fuel required to reach a prescribed velocity in a given engine.

The First, Third Stage Apollo burn

The third stage engine can develop 230,000p of thrust. The total weight of Apollo prior to third burn is 799,645 p.

$$Vg=\frac{(230{,}000p)(32f/s-s)(639.2s)}{326{,}669}=14{,}400\text{ f/s and } R=\frac{799{,}645p}{472{,}946p}=1.691 \qquad \ln(1.691)=.5252$$

Viii=(14,400f/s)(.5252)=7563f/s

The above velocity is just 44 f/s faster than the required 'park' velocity at 118 miles above earth. Figure (A) below, shows the approximate positions of the ship at stages I and II after launch. Figure (B), looking down on earth from above the N Pole). The ship leaves earth at a point which is perpendicular to an axis through the equator.

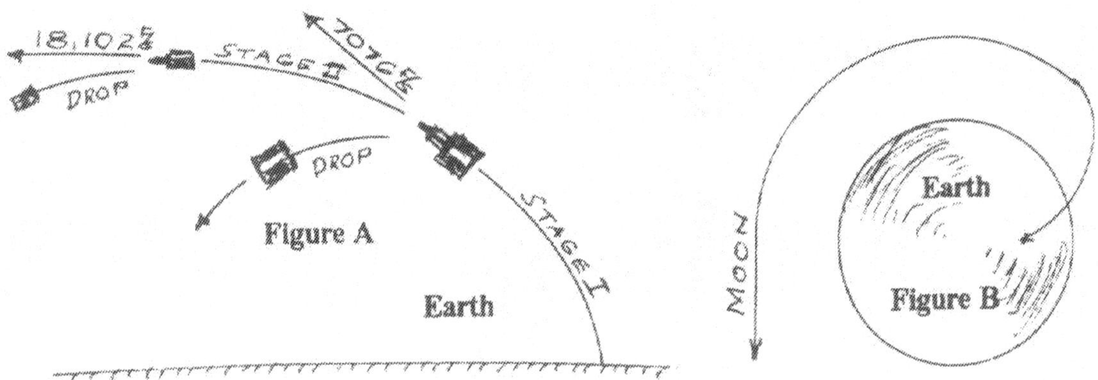

Figure A
Figure B

The Velocity of Escape from Earth

The Apollo and crew (you are the crew) have made 0.5 orbits about earth. Very soon now NASA will fire the third stage, for the second and final time, sending you on your way to the moon. The velocity of escape from earth to the moon is 35,570 ft/sec or 24,252 mi/hr. This number is less than that for escape to the distant planets. Refer tot eh data chart on page (107) last figure lower right, for your escape velocity to the moon and to the NASA announcement, lower right, for the correct velocity of the original Apollo Crew. Using this figure, your ship must gain an additional 9.949 ft/sec.

ΔV=35,570 f/s – 25621 f/s = 9.949 ft/sec

In order to achieve this velocity, your ship will expend 300,000 additional pounds of fuel. After final third stage burn, the tank and remaining fuel will be jettisoned, and you are on your way to the moon. The final calculations are as follows:

$$T=\frac{(Wf)(450s)}{F}=\frac{300{,}000px450s}{230{,}000p}=587\,\text{sec}onds$$

NASA: (July 16th, Translunar injection, at cut off, velocity is 35,570 ft/sec, at 177 nautical miles)

$$Vg = \frac{(F)(g)(T)}{Wf} = \frac{230{,}000px32f/s - sx587s}{300{,}000p} = 14{,}400f/s$$

$$R = \frac{Wt}{Wb} = \frac{472{,}946p}{172{,}946} = 2.73 \rightarrow In(2.73) = 1.01$$

V=(Vg)(1.01) – (Gx')(T)

V=(14,400 f/s)(1.01-(7.38f/s)(587s)=14,544f/s – 4332 f/s = 10,212 f/s

Vesc = Vorb + V = 25,621 f/s + 10,212 f/s = 35,833 ft/sec

(You have no ready way of determining the value of Gx' in the equation to the left, because you have no way of knowing how far you will be from the center of earth. You will be shown how this number was derived later)

The final velocity above exceeds the required velocity to escape to the moon. You are now on a collision course with the moon. The velocity of your ship will decrease by the second until you cross the zero gravity line. After that you will be accelerated toward the moon by the moon's gravity.

You are now in free space. How much time was required for your ship to travel to final burn and what time does your watch read now. It seems like a long time ago that your ship blasted away from the pad at 9:32 AM, Eastern Daylight Savings Time.

Stage I + Stage II + Stage III(1) + Stage III(2) =
190s + 381s + 639s + 587s = .499 hours
.499 + orbit time(0.5 orbits x 1.442 hr/orb)=.499 hr + .721 hr = 1.22 hours
Your present time reads 10:45 AM

```
5 LPRINT "THIRD FIRE"
10 LPRINT "APOLLO – Fourth and Final Earth Firing"
20 LPRINT "W = Fuel wt. (Pounds)"
30 LPRINT "F = Thrust (Pounds)"
40 LPRINT "T = Time (Seconds)"
50 LPRINT "I = Impulse (Seconds)"
60 LPRINT "G = Gravity (ft/sec-sec)"
70 LPRINT "C = Veelocity of gas flow (ft/sec)"
80 LPRINT "V = Ships third stage velocity (ft/sec)"
90 LPRINT "M = Ships pre-burn wt. (Pounds)"
100 LPRINT "A = Angle (Radians)"
110 LPRINT "B = Ships final second stage velocity (ft/sec)"
120 LPRINT "Z = Ships final resultant velocity (ft/sec)"
130 W = 300000!
140 F = 230000!
150 I = 450
160 G = 23.58
170 A = 1.3936
180 M = 472946!
190 B = 25621
200 T = (W * I)/F
210 C = (F * T * 32)/W
215 R = M/(M – W)
220 V = (C * LOG(R) – (COS(1.3936) * G * T))
```

```
230 Z = V + B
240 LPRINT" W   F   I   G   M"
250 LPRINT W, F, I, G, M
260 LPRINT" T
270 LPRINT T, C, V, B, Z
280 STOP
```

Third Fire
APOLLO – Fourth and Final Earth Firing
W = Fuel wt. (Pounds)
F – Thrust (Pounds)
T = Time (Seconds)
I – Impulse (Seconds)
G = Gravity (ft/sec-sec)
C = Velocity of gas flow (ft/sec)
V = Ships third stage velocity (ft/sec)
M = Ships pre-burn wt. (Pounds)
A = Angle (Radians)
B = Ships final second stage velocity (ft/sec)
Z = Ships final resultant velocity (ft/sec)

W	F	I	G	M
300000	230000	450	23.58	472946
T	C	V	B	Z
586.9566	14400	12046.77	25621	37667.77

FINAL PRE-CALCULATIONS – FREE SPACE TRAVEL

Looking at the diagram below, you can see that during the seconds the ship accelerates along its path of intersection with the moon, it has only begun to move away from the earth. What happens to Apollo during this period of time, determines whether or not the ship will reach the orbit of the moon. The lines drawn from the center of earth to ship's position represent a retrogression of angles form 90° to 20°. Only the first line is perpendicular. Below is a table of data and calculations used to make up the graph. This is the method used to determine the 7.38 f/s-s gravity calculation used in the determination of final velocity on page (106).

Gravity Data and Calculations					
Angles		Distance	Gravity(G)		Gravity(Gx)
90°	4118mi/Sin90°	4118 mi	30.19f/s^2	G x Cos 90°	0 f/s^2
80°	4118mi/Sin80°	4181 mi	29.28	G x Cos 80°	5.08
70°	4118mi/Sin70°	4382 mi	26.66	G x Cos 70°	9.11
60°	4118mi/Sin60°	4755 mi	22.64	G x Cos 60°	11.32
50°	4118mi/Sin50°	5375 mi	17.72	G x Cos 50°	11.39
40°	4118mi/Sin40°	6406 mi	12.48	G x Cos 40°	9.56
30°	4118mi/Sin30°	8238 mi	7.54	G x Cos 30°	6.53
20°	4118mi/Sin20°	12040 mi	3.31	G x Cos 20°	3.31

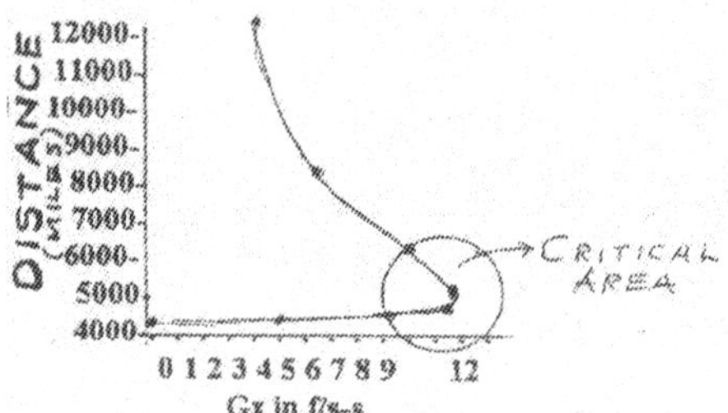

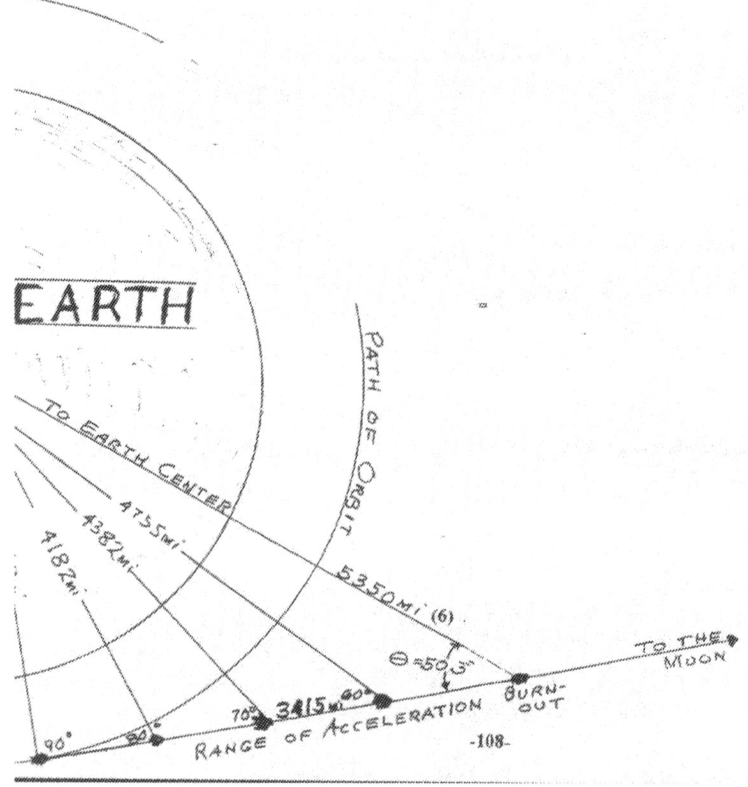

Look at the graph of the above material and you will see that Gx, effective gravity, increases to 11.39 f/s-s and then decreases. 7.38 f/s-s is the average gravity through 50 degrees of angle. After 50 gravity decreases very fast and need not be considered effective in slowing the ship.

Exercise 3

Self Test

Make the following calculations (most of the answers can be found on this sheet of paper). Submit your work to your instructor.

(1) Calculate the subtended angle at the point of final burn.

(2) Use the range acceleration line to calculate the distance from final burn to center of earth.

(3) Calculate gravity at initial fire.

(4) Calculate gravity at final burnout.

(5) Calculate the acceleration (A) of the rocket.

(6) Prove the range accel. Distance to be 3415 mi.

Exercise 33-B

Self Test

You are in a space ship orbiting earth at 300 miles above its surface. The following information is correct about your ship: I = 450s; F = 350,000p; Wt = 500,000p; Wf = 325,000p; Gx = 6.38 f/s-s (Remember that this number is derived using the Cos θ). You may use all of the fuel to make this burn. Make the following calculations and show all pertinent data in a table along the right side of your paper.

(1) Calculate orbital velocity at 300 miles above earth.
(2) Calculate gravity at orbit.
(3) Calculate escape velocity from earth.
(4) What additional velocity (ΔV) do you need to escape earth's gravity?
(5) Calculate the time of burn of yoru ship.
(6) Calculate (Vex), the velocity of gas ejection.
(7) Calculate (R), the burn ratio of the ship, and (In R).
(8) Calculate the final velocity of the ship after burn.
(9) Calculate the acceleration of your ship during burn.

(Assume that your ship left earth orbit along a tangent line perpendicular to earth's surface for the following calculations.)

(10) What total distance in miles did your ship travel along the Range Acceleration Path during burn?
(11) How far in miles is your ship from the surface of earth?
(12) Calculate the angle between the ship's "acceleration path" and a line drawn from the point of burnout to the center of earth.
(13) make a scaled drawing of the above situation, that is similar to the figure on page (108).

Weight of Fuel Equation for a Given Burn

This exercise is based on information provided for the calculation of the amount of fuel needed to achieve a given velocity of orbit. The developed equation below is derived from the stage-burn-velocity equations as follows:

Given I = (F)(T)/Wf; Vg = (F)(G)(T)/Wf; V=(Vg)(InR); R=Wt/(Wt-Wf).

Then if V=Vg(InR) → substitute as follows: (F)(G)(T)/Wf for Vg and (Wt)/(Wt-Wf) for R

Then $V=\frac{(F)x(G)(T)}{(Wf)}xIn\frac{Wt}{Wt-Wf} \rightarrow Arrange I=(F)(T)/Wf\_as$ R/Wf=I/T. Then V $=(I)x(G)(T)xIn\frac{Wt}{Wt-Wf}$ and $V=(I)(G)xIn\frac{Wt}{Wt-Wf}$ then $\frac{V}{(I)(G)}=In\frac{Wt}{Wt-Wf}$

To solve this equation, you must take the ani-log of the left side of the equation.

Example: Calculate the amount of fuel needed to increase the ship's velocity by 5,659 ft/sec.

$V/(I)(G)=\frac{5{,}659f/s}{(450s)(32f/s-s)}=In\frac{799{,}645p}{799{,}645p-Wf} then .393=In\frac{799.645p}{799{,}645p-Wf}$

Next take the anti-log of .393 which equals 1.48 then 1.48 (799,645p-Wf) = 799,645p

1,183,475p – (Wf) = 799,645p

-1.48(Wf) = -383,830p

Wf = 259,344p

Exercise 34

Self Test

Show all required work with the correct units.

Problem (1) Copy the derivation of the above equation and turn this in as the first part of your assignment.

Problem (2) Use the same weight of the ship, impulse and gravity, as used in the above example. Calculate the amount of fuel required to increase your ship's velocity by 12,000 ft/sec.

NAVIGATION CHART –Relative Earth-moon Positions at Launch

The drawing below is a reasonably accurate accounting of all of the distances involved in making an acceptable rendezvous with the moon in three days at third stage firing. Notice that the ship must be launched a few minutes before it has traversed all of its parking orbit. The Apollo moves through space for three days as the moon progresses from M-1 to M-4. Each day the moon moves 49,630 miles through its orbit for a total distance of 148,890 miles in three days. The Apollo must engage the moon, at its equator, on the leading side. On the aft side, it could easily miss the moon.

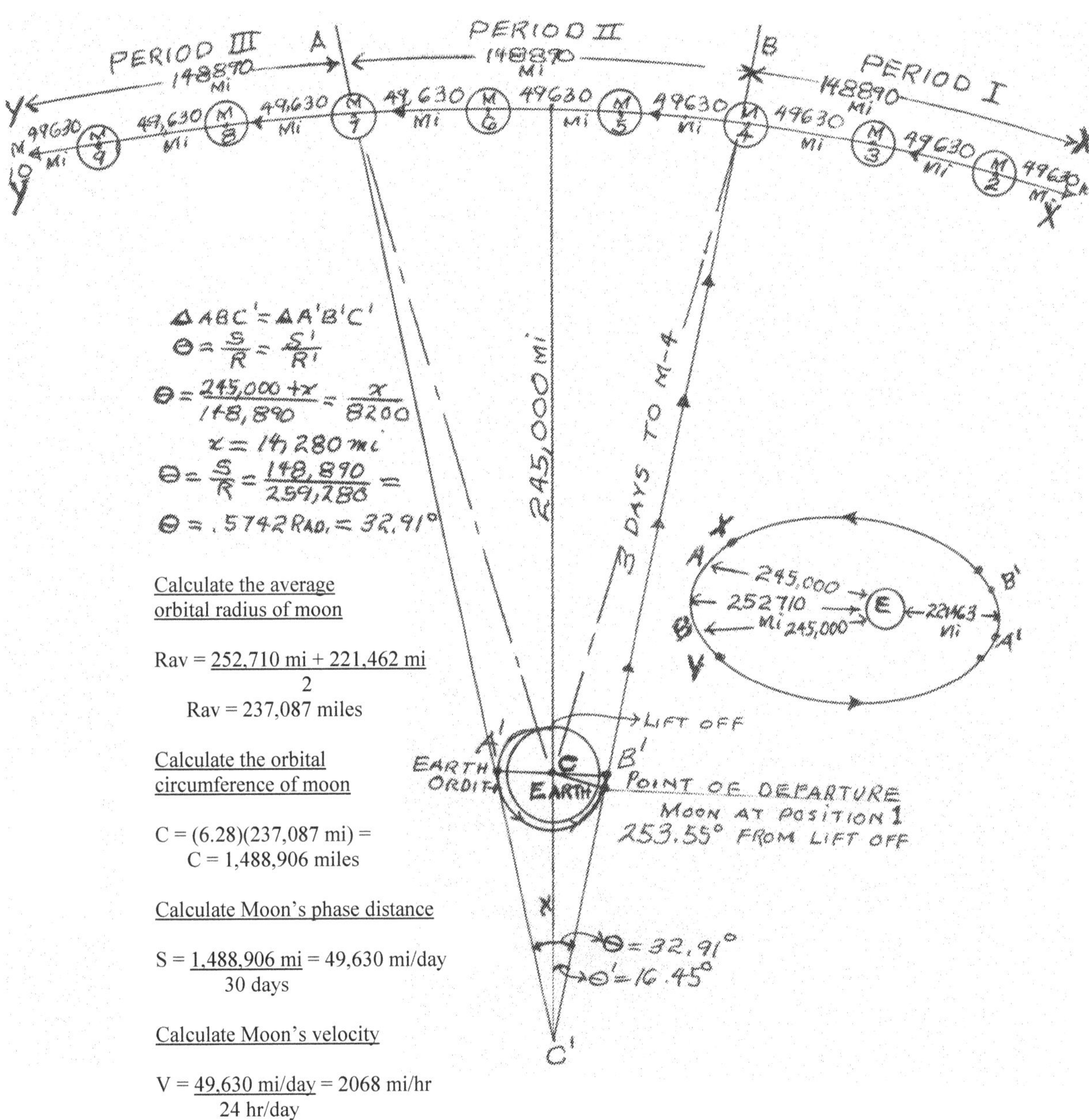

Why the Long Side of the Moon's Orbit

The Apollo distance to the moon is 245,000 miles. Why was the far distance chosen rather than the distance nearest to earth? The nearest distance is the perigee which is only 221,463 miles. It would take much less time to get there than to travel the additional 23,500 miles into the far area of the apogee. This defies logic! The shortest distance between two points is nearly always the best way to get where you are going. But not in this case. At B'A' (small drawing), the moon is much faster than at AB. For this reason, it is easier to miss the moon. The best shot is at AB, and why the end of the apogee? Notice that the moon is traveling away from earth at (A) which makes it easier to catch. Also, observe that the moon is beginning to move toward earth at B. This makes it possible to use a portion of the moon's velocity to propel the ship back to earth when it escapes the moon's gravity. How would you calculate the moon's velocity at apogee and prigee? Why not use Kepler's Law to do this. First recall that the Rav of the moon is 237, 087 miles and calculate the average velocity of the moon assuming an orbital circle.

V.m = (Cmn) / (Pmn) = (6.28)(237, 087mi) / (30 days) = 49,630 mi/day = 2068 m/hr

Next use Keplers Third Law or $P^2 \alpha R^3$ to calculate the moon's velocity at full at apogee and perigee.

$$\frac{(Pxy)^2}{(Pr)^2} = \frac{(Rxy)^3}{(Rr)^3} \rightarrow Pxy = \frac{(252{,}710mi / orb)^{3/2}}{(237{,}087mi)^{3/2}} = (30d)(1.1 / orb) = 33days / orb \text{ at apogee}$$

$$\frac{(Px'y')^2}{(Pr)^2} = \frac{(Rx'y')^3}{(Rr)^3} \rightarrow Px'y' = \frac{(30d)(221{,}463mi / orb)^{3/2}}{(237{,}087mi)^{3/2}} = (30d)(.903 / orb) = 27.09d / orb \text{ at perigee}$$

From the above, you can see that the 'apogee' orbit would take about three days longer for the moon to make its trip around earth. The 'perigee' orbit would take about three days less. The average of these two values is about 30 d/orb.

Exercise 35
Self Test

The following questions are about the Navigation Chart on the previous page.

(1) How far is the distance from earth to the moon?
(2) If the moon's gravitational field begins 30,000 miles from the center of the moon, how far will it be to zero gravity for either planet? (Assume the center of earth.)
(3) How many miles will the moon move through its orbit from the time of third stage firing until zero gravity is reached? (The answer is not the total three day distance.)
(4) How far is it between successive positions of the moon?
(5) What distance does the moon travel in one period.
(6) When the moon is in position M2, what distance has Apollo traveled from its point of departure from earth orbit? (You may assume a fractional part of the distance from earth to moon.)
(7) What is the significance of the quotient of the moon's orbital circumference and the distance the moon travels in 24 hours?
(8) Can you foresee any problem related to the return trip that was not a problem on the trip to the moon? What additional factor must be taken into consideration on the return tip to earth?
(9) How far does the moon travel in 30 days? Is this equal to or greater than the circumference of its orbit?
(10) What must be the average velocity of Apollo if it is to arrive at the moon in 3 days?

Exercise 36

Self Test

The moon Sram orbits a planet with an apogee of 25,000 miles and a perigee of 200,000 miles. If its period of revolution is 40 days, calculate and/or answer the following:

1. Make a drawing of the orbit of Sram about its planet and label the apogee and perigee showing distances.
2. At which end, apogee or perigee, will its velocity be greatest?
3a. Calculate its average radius of orbit, and (b) calculate the circumference of its orbit.
4a. Calculate its average velocity in mi/day, and (b) in mi/hr.
5a. Calculate the velocity of the moon in mi/day, and (b) in mi/hr at the apogee.
6a. Calculate the velocity of the moon in mi/day, and (b) in mi/hr at the perigee.
7. Which part of the targeted moon is easiest to catch and why?
8. Is the relationship between velocity and its associated radius direct or inverse as applied in Kepler's Law?

Chapter IX

Free Space Maneuvers

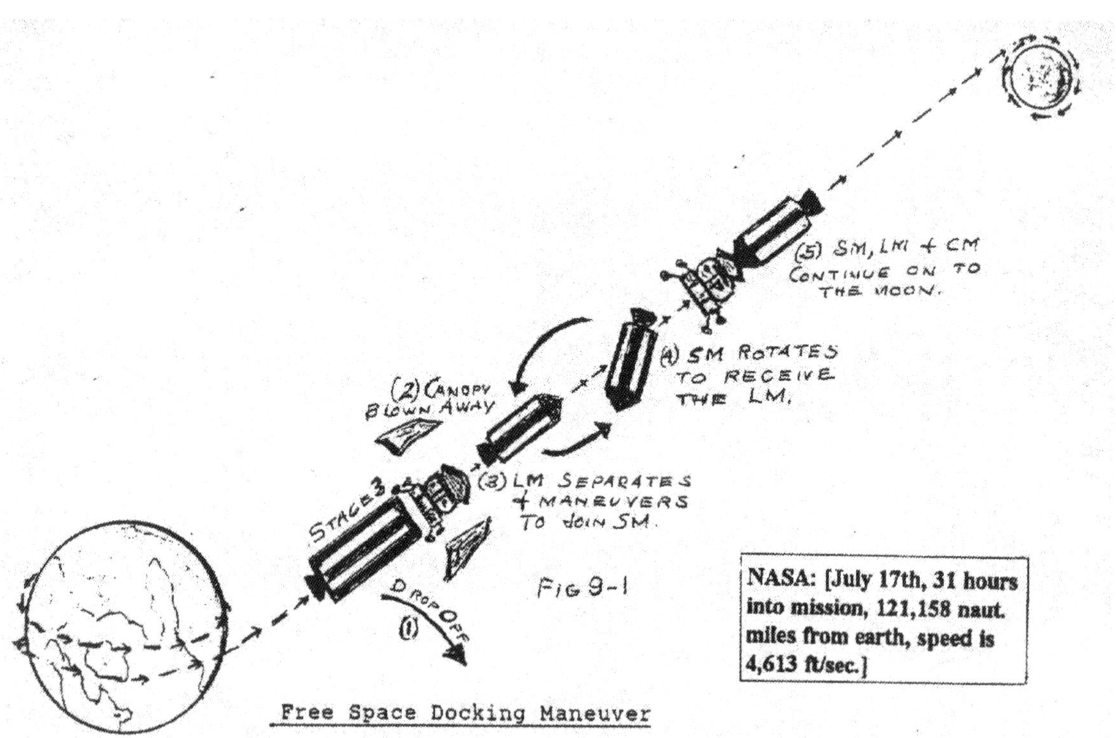

Free Space Docking Maneuver

Long before you boarded the Apollo, you prepared for the next phase of the trip to the moon. Now is the most critical time of all. You are about to begin changing the order of the remaining three modules in which you ride, still attached to the third stage of the rocket. The order of sequence of the LM, the Service module and the Command module must be changed prior to entering orbit. The problem is that the LM is attached to the service module motor. Firing the service module motor would cause the entire assembly to explode. Because of this problem, a critical maneuver need be performed at this time. With the third stage still connected, the shroud must be blown away and the multiple attaching linkage removed. After this is done, the module is reversed end to end with its nose attached to the LM as shown in the drawing. The intricate docking procedure is done and the third stage is separated and set adrift in space. For the first time, it is possible to move back and forth through all three sections as it continues its drift toward the moon. The engine of the service module is now in position to fire and to slow down the ship as it enters moon orbit.

All members of the crew were aware that any serious problems in creating the new linkage would mean an end to the trip to the moon. This was another critical part of the trip to the moon that was executed faultlessly.

Notice that the earth is fading away and the moon appears bigger and closer than ever.

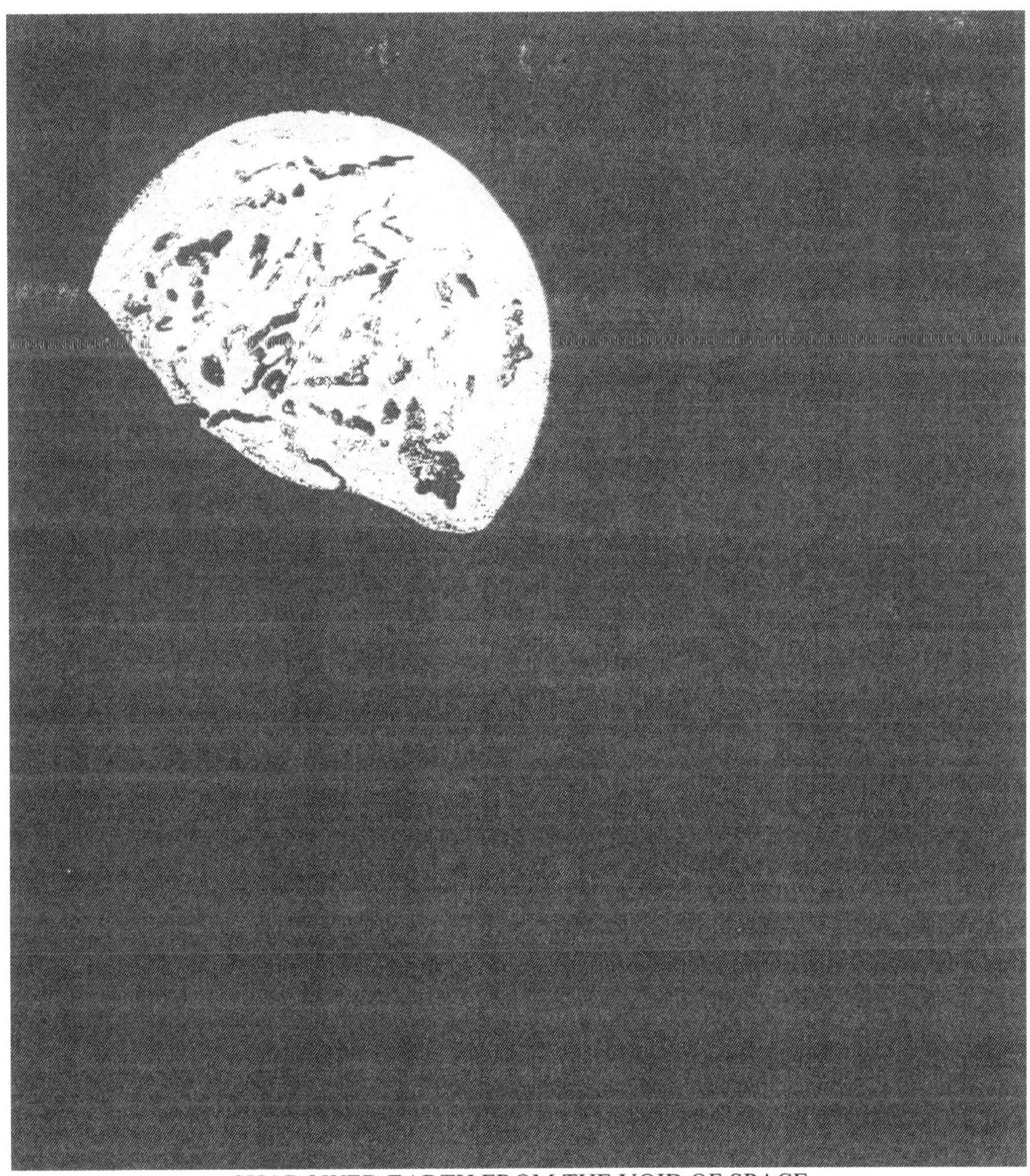

SHADOWED EARTH FROM THE VOID OF SPACE

Blue and white earth as seen from space. Clearly defined is the sun's point of deliniation. The earth has become smaller and smaller becoming a true isolated spheroid in space. The dark, ominous moon grows ever larger, its cold, unknown surface features are now more prominent.

FREE-SPACE-Earth to Moon

The following program will take your ship from "burnout" at earth to zero gravity near the moon.

```
10 LPRINT "FRESPACE-TO THE ZERO GRAVITY INTERFACE"
20 LPRINT "APOLLO- First Manned Flight to the Moon"
30 LPRINT "VELOCITY    DISTANCE    TIME    GRAVITY"
```

```
40 A = 35865! 'Escape velocity form earth
50 T = 1
60 S = 141 'Distance from the surface of earth
70 G = 29.96 'Gravity at 141 miles from earth
80 P = 0
85 N = 0
90 LPRINT V,S,N,G 'First print statement for velocity, distance
100 C = G * 1 'Changes gravity to velocity, time and gravity
102 GOTO 110
105 C = ((G+X)/2)=1 'Averages gravity per second
110 V=A-C 'Subtracts gravity from ship's velocity
115 X=G
120 D=(1*V/5280)+P+S 'Calculates the distance of ship
130 R = 4000+D
140 G=32*(4000/R)*2 'Calculates the new per-secodn gravity
145 N=N+1 'This is the time adder
160 P=D
170 S=0
180 A=V
190 IF D >215000! THEN GOTO 220 'The end of the program statement
210 GOTO 105
220 LPRINT V,D,N,G
230 STOP
```

FRESPACE-TO THE ZERO GRAVITY INTERFACE			
APOLLO-FIRST MANNED FLIGHT TO THE MOON			
VELOCITY	DISTANCE	TIME	GRAVITY
35865	141	0	29.96
1794.077	215000.3	247582	1.067531E-02

The moon is 30,000 miles away from a final destination with Apollo. The moon's velocity is 2063 mph, and it moves a distance about equal to its diameter each hour. When your ship crossed the gravitational interface, its velocity was 1794 mi/hr. It is now being accelerated by the moon's gravity which is 5.33 ft/s-s at the moon's surface. Using average gravity and the final velocity of Apollo at zero moon gravity, a simple algebraic, quadratic equation will tell you that you will crash on the moon within 2.79 hours if no other forces intervene to prevent this from happening.

Now you must write a Basic program that will track the ship into parking orbit at 70 mils above the surface of the moon, if you are to prevent your crashing on its surface.

Exercise 37

Self Test

Use the "Moon Interface" drawing as a reference to the problems below.

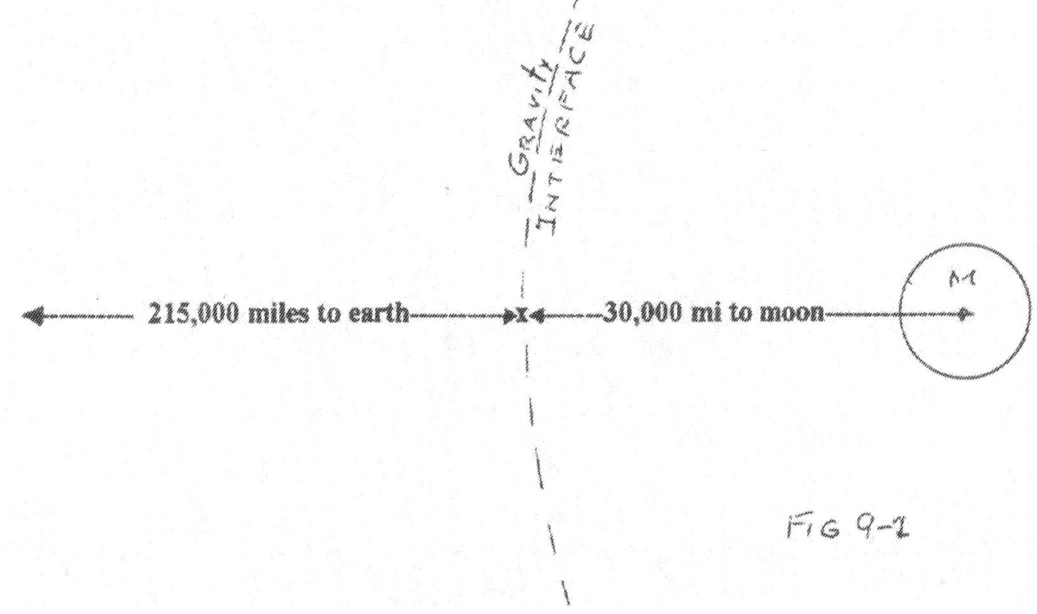

(1) Assume that Apollo arrives at gravity interface (zero gravity) with a very small positive velocity. If it is 30,000 miles from this point (x) above, to the center of the moon, how many hours would pass before the ship would crash on the surface of the moon? (Surface gravity is 5.33 f/s-s and the moon's radius is 1087 miles.)

(2) Using the "Frespace" data printout on page (110), you can tell that Apollo entered the gravity interface at 1794 f/s or 1223 mi/hr. Write and solve a quadratic equation that will prove or disprove that Apollo would crash on the moon in 2.76 hours if retro-engines did not fire. Hint: first calculate average gravity. You are aware that a Basic program would provide a more accurate answer than would an algebraic solution to the problem. EXTRA CREDIT to you if you can write a Basic program, incrementing the time by one second, beginning at zero moon gravity, that would provide for a more correct answer to this problem. (Submit your program, even if it does not run correctly—or do as your instructor tells you to do.)

> NASA: (July 18th, 48 hours and 58 minutes into flight, 146,300 nautical miles from earth, velocity 3,917 ft/sec...Lunar landing 61 hours and 48 min. away.)

Looking Back to Earth, Then at the Awesome Moon

Now the earth looks like a large, peaceful, blue and white beach ball. But, the moon is coming straight at you, growing ever larger, dark and ominous. You can't avoid it. It is there. NASA put you exactly where you should be. You must either crash into it or make orbit around it. You have few options and less time. The picture on the opposite page tells you that others have been here to look over its surface. Mare Tranquillitatis is plainly visible and that is where you will land if you can.

If Apollo 11 is to land on the southwest corner of the Sea of Tranquility (Mare Tranquillitatis), you need to give careful attention to the best possible approach trajectory of your ship that will make for easiest access to the chosen landing place. Bear in mind the place of landing should be as near to being under the Command Module as possible. This will make it much easier for the landing crew to take off and re-join the Command Module, as a point directly beneath its parking orbit is the shortest distance between the two objects.

As you recall, the Sea of Tranquility was chosen because of its unusually large span of flat surface located near the equatorial center of the moon. Notice that Mare Serenitatis is about the same size as the Sea of Tranquility but is better defined. Measure the diameter of the broad plain of Serenitatis from left to right. Use 2172 miles as the average diameter of the moon and calculate its approximate surface area. Did you get 138, 474 square miles? You should easily be able to land your ship in an area that size.

Looking at the pictures, you can see hundreds of craters (actually there are millions of craters on the surface of the moon). All are the mark of some meteor that came from outer space when the moon was newly formed and its surface thin and hot. There would be much danger for a crew landing in some unchartered area. Should the landing craft tilt or fall over, it would be impossible for the crew to properly orient the ship on take off. Any crash the crew survived, would be the end of the crew in just a few hours. The objective of this trip is to get all of you back to earth in good health.

The series of pictures on the next page are designed to show the relative position's of the ship and the moon as the ship enters zero gravity. The ship is arriving at zero gravity after 68.8 hours of space travel. You have arrived short of three days and several hours ahead of the moon as planned. If all goes according to plan, as shown in the drawing, the Apollo will drop into moon orbit in the proximity of 70 miles above the surface of the moon. In order to stay in orbit, its velocity must be 3653 mi/hr. If you compare the difference between your velocity at zero gravity and that at orbit, the difference is 2430 mi/hr. The trajectory of the ship assures that an average angle of 64 degrees with the vertical will be maintained in order to make up for this difference.

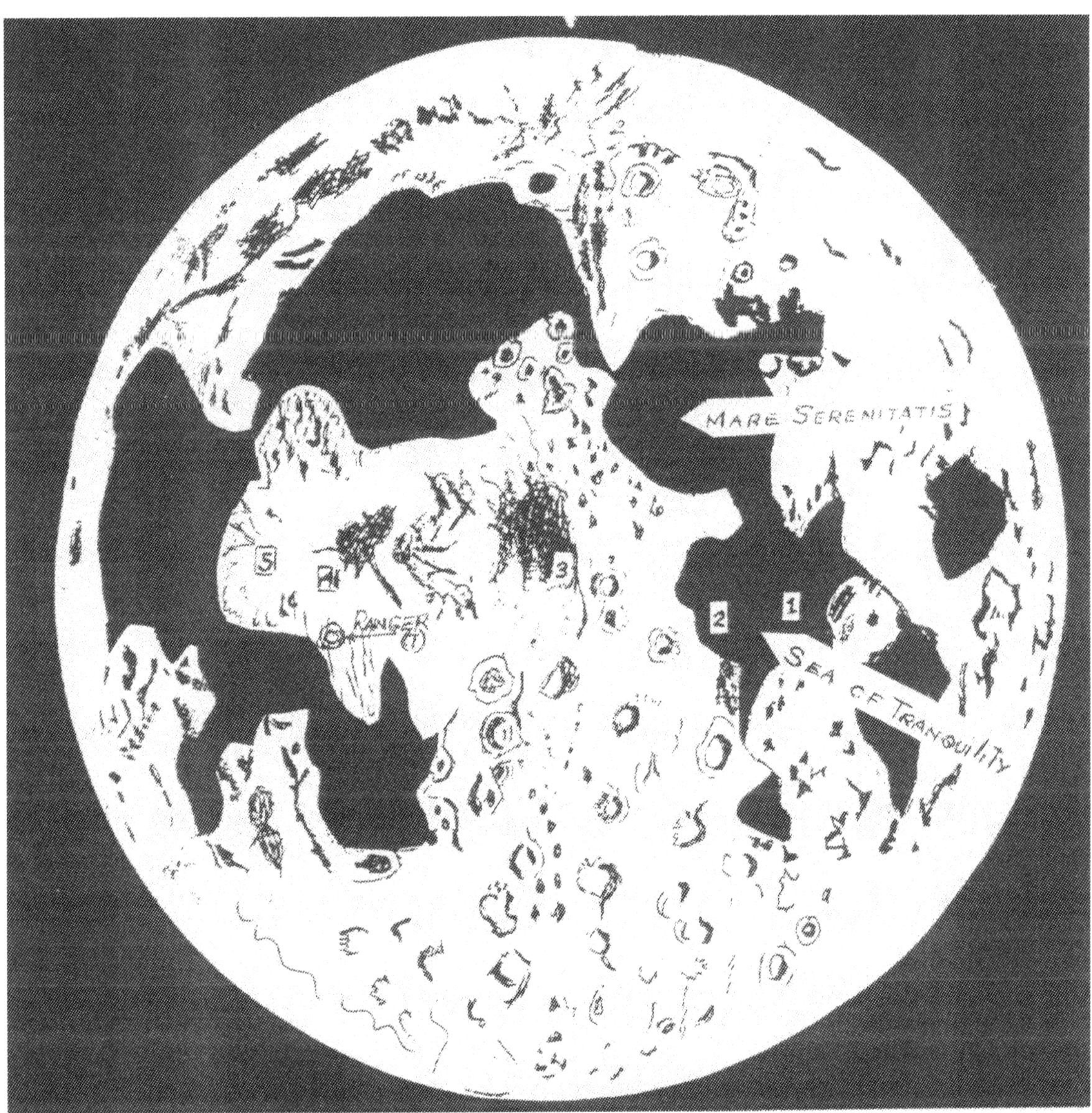

PROPOSED LANDING SITES FOR APOLLO

The first of these was the Ranger Series of probes. These probes were to take television pictures which were to be transmitted back to earth just before the probe crashed on the surface of the moon. After many failures, on July 31, 1964, Ranger VII of the B series, transmitted 4000 photographs of the lunar surface showing craters as small as 2 ½ feet in diameter. It crashed in Mare Cognitum and is designated Ranger (7) in the picture above. In 1965, Ranger VIII photographed the Sea of Tranquility which became the chosen landing site of Apollo 11. As NASA's rockets improved, the entire Surveyor series of probes followed in 1966 and 1967. In all, five sites for Apollo landings were proposed. These are shown as white dots in the above picture and designated 1-5. Note that all are along the equator of the moon.

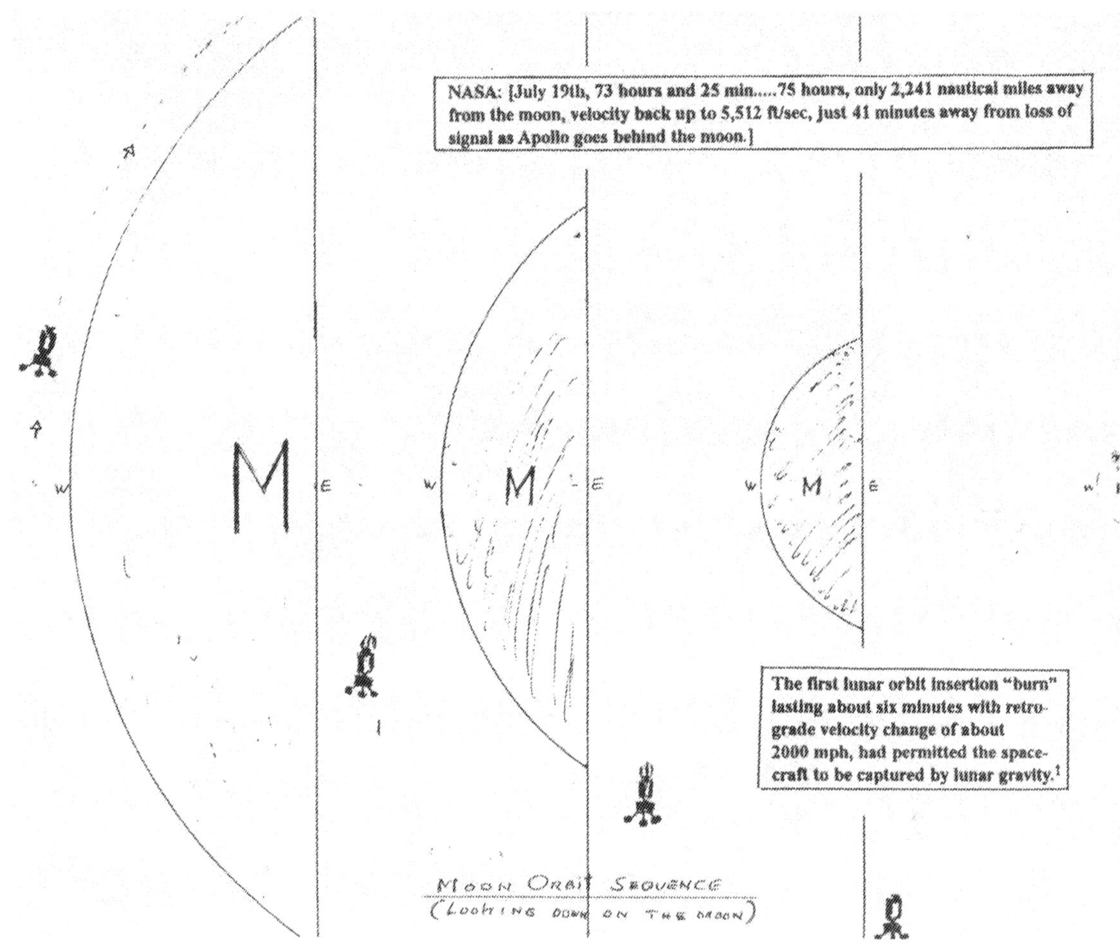
NASA: [July 19th, 73 hours and 25 min.....75 hours, only 2,241 nautical miles away from the moon, velocity back up to 5,512 ft/sec, just 41 minutes away from loss of signal as Apollo goes behind the moon.]
W
M
E
W
M
E
W
M
E
The first lunar orbit insertion "burn" lasting about six minutes with retrograde velocity change of about 2000 mph, had permitted the spacecraft to be captured by lunar gravity.[1]
MOON ORBIT SEQUENCE
(LOOKING DOWN ON THE MOON)

What Remains is Apollo 11

Now is the time to take stock of your ship. The drawing to the right represents the three remaining sections of Apollo after the first three stages are gone. The entire assembly weights 100,000 pounds and is made up of the following components:

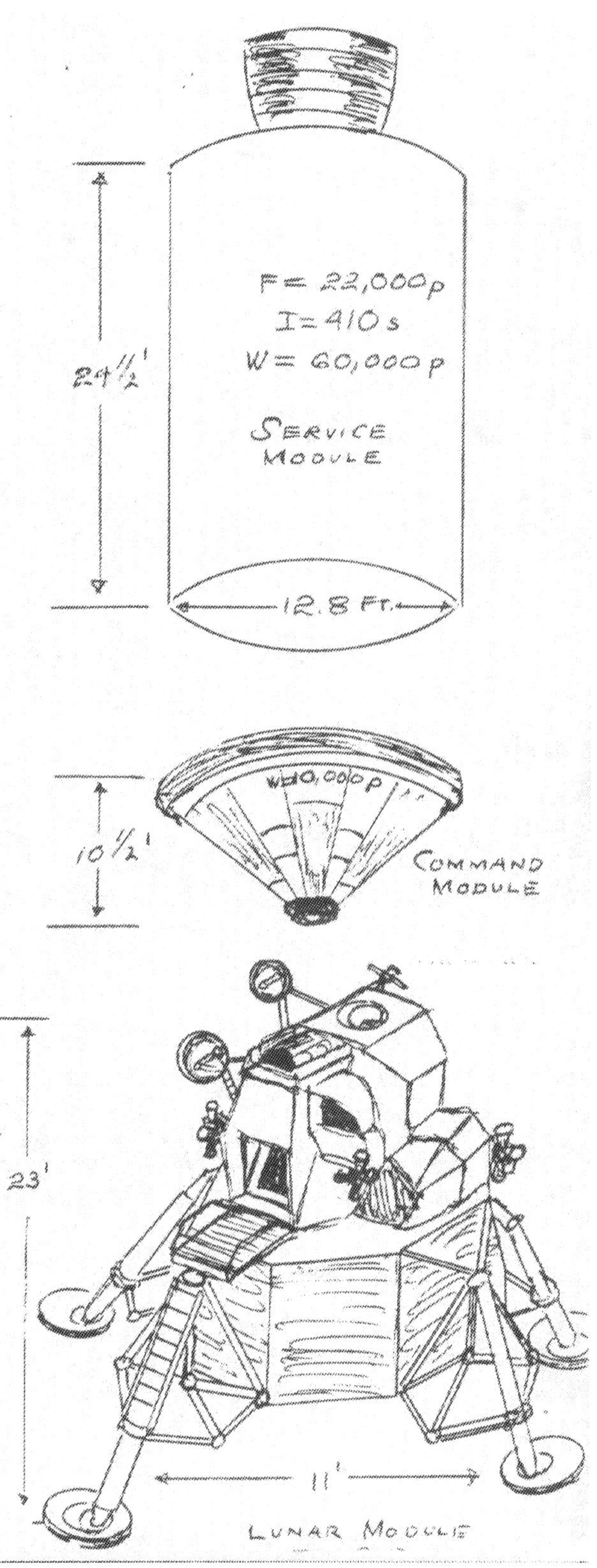

The Service module, is the leading portion of the ship which weighs 60,000 pounds and carries mostly fuel with an attached engine. It will be retro fired to slow down the ship as it enters orbit around the moon.

The Command module is the re-entry ship that will take you back to earth and land in the ocean. It weighs 10,000 pounds. It is designed to protect you from the heat of re-entry.

The Lunar module is the landing module that will place you on the surface of the moon. After that, the landing pod will be left behind and it will place you back in orbit to re-unite with the Service portion of the ship. The LM weighs 30,000 pounds. Its engine is capable of 35,000 pounds of thrust.

Chapter X

The Moon Has Arrived

Making First Orbit Around the Moon

At first the moon appeared far away but moving toward us. With the moon so distant from us, we were more concerned that we may not crash on the moon but that we could miss it and continue on out into space.

Collins wrote in his log, "The moon I have known all my life, that two-dimensional, small yellow disk in the sky, has gone away somewhere, to be replaced by the most awesome sphere I have never seen…It is huge, completely filling our window." In addition, he wrote that he did not like the looks of the moonscape, he found it "distinctly forbidding" and "stark and barren."[2]

From First Orbit to 60 Miles (Lunar Orbit Insertion Two)

Apollo entered orbit within 19.54 hours after passing Gravity Zero. Its final velocity at 70 miles above the moon was slightly in excess of the required velocity to orbit. The ship engine had to be fired to slow it down to 5,358 ft/sec. This amounted to a nominal velocity reduction of 54 ft/sec and used very little fuel. The average angle of entry was 64 degrees with a vertical drawn to the center of the moon. NASA has done a great job and so has the crew. That was a magnificent entry into orbit-not always to be executed so perfectly. The Apollo orbit will next be adjusted to 60 miles above the moon. This will be routinely done with no further calculation necessary to support the maneuver.

```
10 PRINT "MOON 1 ZERO GRAVITY TO MOON ORBIT AT 70 MILES"
20 LPRINT "APOLLO-First Manned Flight to the Moon"
30 LPRINT "VELOCITY   DISTANCE   TIME"
40 A = 1794 'Ship's velocity at gravity interface
50 T = 1
60 S = 30000 'Distance to the center of the moon
70 G = 0
75 V = 1794
77 O = 1.112 'Average angle in radians
80 P = 28844 'Distance from zero gravity to orbit position
85 N = 0
90 LPRINT V,S,N
100 C=(G*1)*Cos(θ) 'Velocity angular correction
110 V=A+C 'Per second interval velocity
120 D=P-(1*V/5280) 'Per second distance from the moon
130 R = 1086 + D
140 G=5.33*(1086/R)^2 'Per second correction in gravity
145 N=N+1 'Time counter in seconds
160 P=D
170 S=O
180 A=V
190 IF D<) THEN GOTO 220 'Ship is positioned 70 miles from the moon
210 GOTO 100
```

220 LPRINT V,D,N
230 STOP
Printout for Basic Program Moon (1):

Moon 1		
APOLLO-First Manned Flight to the Moon		
Velocity	Distance	Time
1794	30000	0
5412	-.0031	70350

Apollo, now in first orbit, will begin to circle the moon in about 1.967 hours. It will not complete this orbit at this level, as NASA will reduce the speed of the ship for its insertion into final orbit at 60 miles above the moon. This will be somewhat like coming down a roller coaster from its high point and then going up again to a lesser high point. In order to do this, the ship will slow down in order to let gravity begin to pull it toward the center of the moon. As soon as it is sixty miles above the surface of the moon, NASA fires its engines to accelerate it to 3,669 mph. As this velocity is only 11 mph faster than at 70 miles, this is but a minor adjustment.

It is now late on the fourth day of Apollo's trip. This has been a very long four days and you need to rest in preparation for the biggest event yet to come. Tomorrow, you walk on the moon. A final warning! All did not go well for the original crew when they made their final descent to the surface of the moon. They very nearly did not make it.

The Fifth Day

On awakening, Armstrong and Aldrin donned pressure suits and crawled through the 30 inch diameter connecting tunnel into the LM, now called the Eagle. The LM had no seats...the Eagle unlocked form the Columbia setting it adrift in space. Collins, alone in the Columbia, circled the LM inspecting it for any possible damage. Collins would be alone for the next 24 hours. If the Eagle crashed, he was to return to earth alone, as there was no possible way he could rescue them.[3]

There is only room for two of you, the rest must remain behind. Put on your pressure suits and lock the hatch as soon as you are located in the LM. Before disconnect, the service ship will begin a quarter rotation which will place the LM in a position between the moon and the service ship. When the LM position is right, check out all systems for "go," then press the release switch when you are ready for disconnect. Upon release from the mother ship, the Command Crew will check you over to be certain there has been no damage to the Eagle during the transposition series of maneuvers. As soon as you are clear, you must orient the LM in order to place the motor in the correct position for retro firing the LM. When this series of maneuvers is complete, NASA will begin a series of reverse thrusts that will take you to 25,000 feet above the surface of the moon.

> Columbia (Collins): Separation performed as scheduled...Separation about 1100 feet at the beginning of "descent orbit insertion,"...lunar landing a little more than two hours away...[4]

Descent to 25,000 Feet

The Eagle is powered by a variable 30,000-10,000 pound thrust throttling engine that can be turned on and off hundred of times during descent to the surface of the moon. You will be able to fly the LM as did Commander Armstrong. The engine will be fired at 25,000ft. to slow down its horizontal velocity component and its vertical component of velocity. At that time, you will take over as did Armstrong and the second member will call off velocity, altitude and angle as did Aldrin.

The on-board guidance computer has just retro fired the LM slowing down your ship by 100 mph. Your velocity is now 3569 mph and you are dropping toward the surface of the moon very fast. Bear in mind that the gravitational force on your ship is partly offset by the centrifugal force which is due to the ships velocity as it travels a quasi-orbital path about the center of the moon. As long as you do not fire your engines, you will continue downward due to the force of gravity, but your horizontal velocity component will remain the same, ie. 3569 mph, until your engines fire again at 25,000 feet. The downward velocity of the ship will determine the length of time over which the horizontal component of velocity acts and therefore the horizontal distance traveled by the ship.

```
10 LPRINT "MOON 2B APOLLO-FIRST MOON LANDING"
20 LPRITN "Sixty mile orbit to 25,000 feet"
30 LPRINT"VELOCITY   DISTANCE   TIME   DG"
40 VX=5258 'Horizontal component of velocity in (f/s-s).
50 R=1146 'Radius of the moon in miles
60 DG=.22 'Delta G=Gravity-A(Radial accel.)
70 VY=0 'Vertical velocity component (f/s)
80 N=0 'Counter in time units of seconds.
90 LPRINT VY, R, N, DG
100 VY=VY+DG
110 D=R-(VY/5280) 'Vertical distance in miles.
120 A=(5258^2)/(D=5280) 'Radial acceleration (F/s-s)
130 DG=6.33=(1086/D)^2-A
135 R=D
140 N=N+1
150 IF D<1090.7 THEN GOTO 170
160 GOTO 100
170 LPRINT"VY! "VY!, "D! "D!
180 STOP
```

The Basic program to the left is designed to calculate the per second change in vertical velocity, in radial acceleration and in gravity, and to calculate the net gravitational effect per second. In the program, item (130) calculates the new radial acceleration and item (140 does the net gravity calculation.

MOON 2B APOLLO-FIRST MOON LANDING
Sixty mile orbit to 25,000 feet

VELOCITY	DISTANCE	TIME	DG
0	1146	0	.22
VY!	449.6856	D!	1090.641
N	1512	DG!	.4838057

During this part of the descent of Apollo 11, the navigational computer is in control. Later on, Commander Armstrong had to take over the controls of the ship in order not to abort the entire operation. In you case, the computer will continue to function correctly, but you will take over the controls for the final landing sequence.

The velocity value of 449.7 ft/sec is (y) or vertical component of velocity. As there is no atmosphere, the (x) or horizontal component of velocity is unchanged and remains 3569 mph.

Exercise 38

Self Test

(After separation, the LM in descent, before second burn)

(1) Calculate the orbital velocity of the Command Module, at 60 miles above the moon.
(2) What was the LM's velocity after first retrofire?
(3) What vertical distance in miles did the LM fall in reaching 25,000 feet?
(4) When the ship has fallen to 40 miles above the moon's surface, calculate (a) Gravity at that point; (b) Radial acceleration at this elevation; (c) The net resultant affect on the ship of (a,b); (d) Will you feel any body weight?
(5) What horizontal distance did the LM travel before it reached 25,000 feet above the moon? (b) Assuming the orbit is equatorial, what fractional part of one orbit does this distance represent? (See the Basic printout on page (124) for this information.)
(6) At 25,000 feet, what are the Vx and Vy components of the ship's velocity in ft/sec and in mi/hr?
(7) Make a fully labeled vector drawing of the x and y components of velocity. (b) Determine the resultant velocity from your drawing. (c) Calculate the resultant velocity and compare this with your scaled determination of this value for numerical error. (d) Calculate the angle of the resultant with the vertical.
(8) What is the significance of the resultant velocity and angle that must be considered when the LM begins descent under second fire? (This requires a written answer.)

Angular Distance on the Moon

(For the next several hours, we will leave our friends suspended at 25,000 feet or 4.73 miles above the moon. During this time you will learn how to keep track of where they are on the moon's surface and when in orbit above the moon.)

Already you know about distance measurement and curved surface measurement on the moon. From your study of geography, you learned about longitudinal and latitudinal line on earth. Cartographers have applied the same concept in mapping the surface of the moon. Reference the map on P-129 and you will see that its surface carries the same mapping configurations as that of earth. Notice the latitudinal lines run N and S and that longitudinal lines run E and W. The primary north-south meridian on the surface of the moon splits the part of the moon that we see into East and West. This is the latitudinal meridian and runs through longitude 0 degrees. This imaginary line runs nearly between Mare Imprium and Mare Serenitatus in the North and along the craters Ptolemaius through Purbach and Walter in the South. You can see where the latitudinal lines are numbered below the equator which runs horizontally across the center of the map section. As the moon is a sphere, it is divided into 360 degrees N to S and E to W. Each grid section is ten degrees on a side. In order to simplify this concept, assume that all numbers begin at the large zero in the center of the map, ie., that numbered latitude lines begin left and right at 0° and count 10°,20°,30°...and that lines above and below 0° count 10°, 20°, 30°...This information can be used to locate places on the moon. For example, the crater Ptolemaius is located at 7°-13° S Lat, and 2° E Long-6° W Long. Your principal landing site is marked with a bulls eye under the second "T" in Tranquillitatis. This is where you are to land the Apollo. The exact coordinates are 2° N Lat. And 34° E Long.

Exercise 39

Self Test

1. Use Lat. And Long. Reference to locate perimeters of the following: (a) Mare Humorum, (b) Mare Tranquillitatis, (c) Mare Nubium, (d) Sinus Iridum.
2. Locate the coordinates of the centers of the following: (a) Crater Schickard, (b) Crater purbach, (c) Crater Atlas, (d) Crater Copernicus.

In completing the above work, you recognize that there is increasing error, due to distortion, as you move toward the top and/or bottom of the map. The radius of the moon at its equator is 1087 mi., but this radius grows smaller as you move toward either pole. For this reason, each longitudinal circumference is smaller than the one before. For future work, note that each new radius changes to become Cos θ x 1087 mi. At 60° Long. The new radius is Cos 60° x 1087 mi.

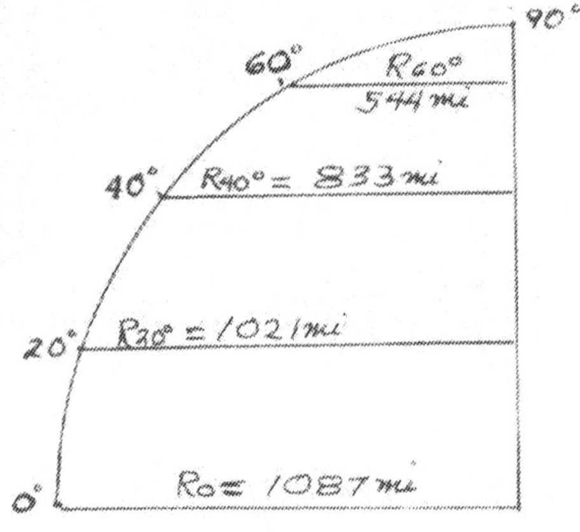

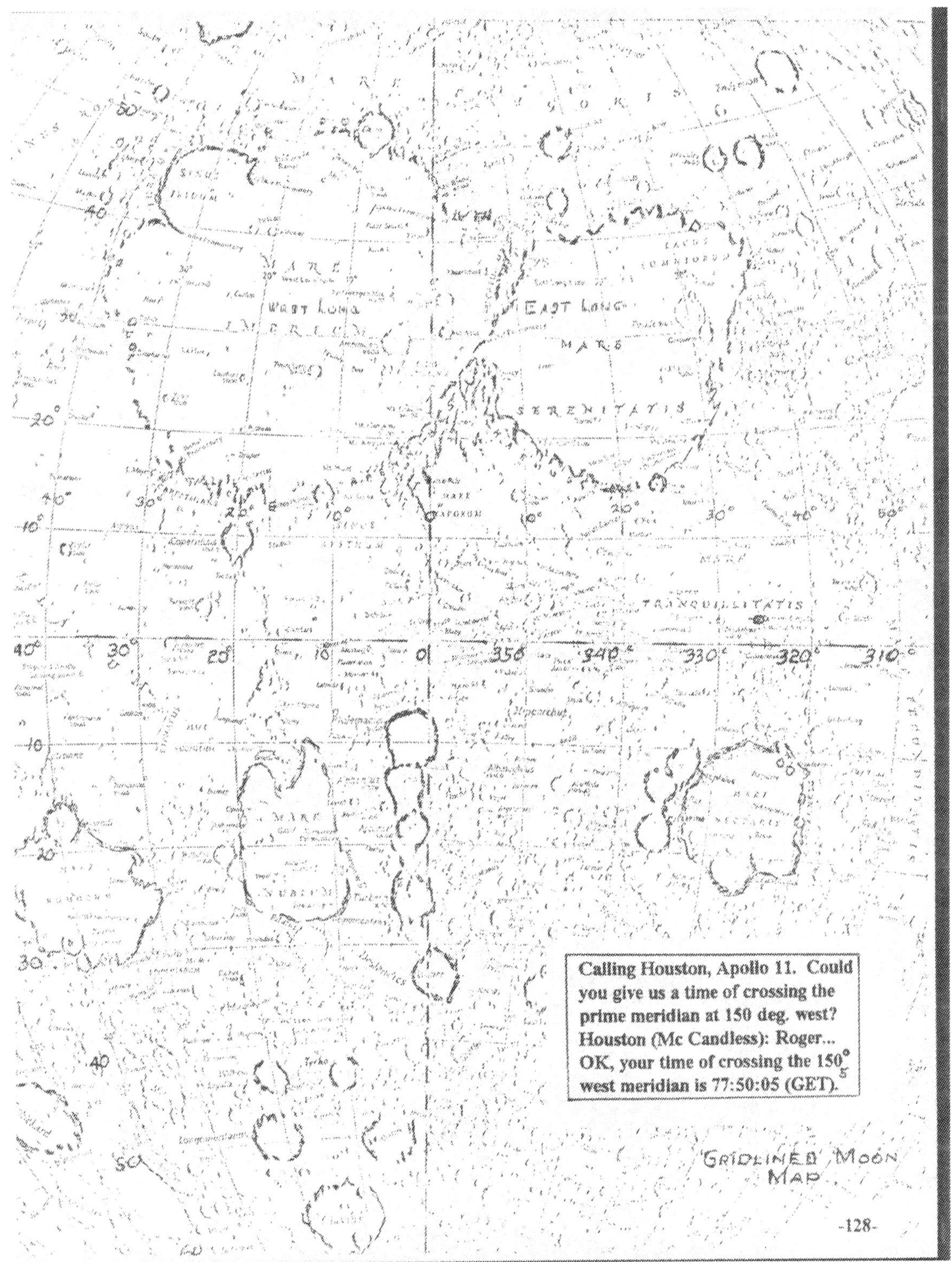

If the moon's equator is divided into 360°, how would you calculate the number of surface miles per degree? This is done by dividing the circumference of the moon by 360°. This value is 18.96 mi/degree. What is this value at 10°? Did you get 18.67 mi/degree? Use this value to calculate the distance across the crater Ptolemaius. Was your answer 97.6 miles? Did you remember to find the circumference of the moon at 10°S. longitude? If you had trouble with this measurement, measure the distance across the grid at ten degrees. Set up a ratio between 10° and this distance. Next measure the distance across the crater and calculate the number of degrees in this distance. Multiply this distance by the circumference ratio at 10°. Your result should be very nearly the same as that above. The mathematics of these calculations is simple as long as the distances are vertical or horizontal.

Exercise 40

Self Test

(Use a chart to display all answers)

(1) Use the above method to find the distances across the following: (a) Ptolemalus and Purbach; (b) Mare Nectaris; (c) Crater Gessendi; (d) Crater Schickard.

(2) Find the distances between the centers of the following: (a) Ptolemalus and Purbach; (b) Willheim and Tycho; (c) Hercules and Atlas.

Using Grid Coordinates to Determine Lunar Positions and Distances.

Now look at the numbers across the middle of the lunar map on P 103. You will be using theses numbers because the face of the moon, where all of the action takes place, ie., the visible portion, is much easier managed with numbered degrees from 0° to 90° and from 270° to 360°. This way, the back of the moon is numbered from 90° to 270°. Also, notice that the degrees are numbered in the direction in which the LM and Service Module are traveling. You can switch to the standard system whenever you wish to do so. How does all of this work? If the Apollo entered lunar orbit at 90°, ie., the left side of the reference map, how far had it gone after passing around the back of the moon to 270? This is one half of a complete orbit or 180. Bear in mind that it orbits the moon at 70 miles.

360° = 6.28 x (70 mi + 1087 mi) = 7266 mi

Distance = 0.5 x 7266 mi = 3633 mi

Another example: If after separating from the LM, the service ship moves into a position at 110°, while the LM has moved to 15° after retro firing, how far apart are they? (Assume both ships to be at 60 miles above the surface of the moon). Was your answer 1900 miles?

What is the Time at 25,000 feet above the moon

You are suspended at 25,000 feet above the moon. What time is it? The time is important because "it" is your common orientation to NASA and to the Command module. The last time you gave any consideration to "time" was just after final third stage burn, as the crew entered free space. That was 10:45 AM, EDST. From this point on, you will operate on Standard Time.

DAY I at 9:45 AM, final burn, 7/16/69

DAY I, II, III, IV on 7/19/69; 247,582 seconds or 68.77 hrs. to Gravity Interface.

Day IV, at 6:31:12 AM; (Gravity Interface to Orbit) 70,350 sec. Or 19.542 hrs.

DAY V, at 2:03:43 AM, on 7/20/69. Allow 30 minutes to change orbit to 60 form 70 miles above the moon.

DAY V, at 12:26:42 PM. The LM disconnected from the CM. How many orbits did Apollo make and where is it at disconnect?

Calculate the time from orbital change to 12:26:42 PM.

Hours = 12.445 – 2.057 = 10.388

This equals 5.3 orbits. It entered "first orbit" at 90 degrees. 0.3 orbits equal (0.3 x 360° = 108 deg.) (Ref. Moon Chart on page (132)). Angular moon position = 198 deg.

The CM was to inspect the LM for damage for one orbit. If there was no damage to the LM, its engines were to retro-fire at a position of 198 degrees, after which it would begin to fall toward the surface of the moon. (Moon chart on p (132.)

There was no damage and the LM engines fired. The LM began to drop toward the surface of the moon. Retro-fire took place at 2:24:18 PM, when the LM was at an angular position of 198 degrees. The velocity of the LM was reduced by 100 f/s to 5,282 ft/sec.

The LM was retro-fired again at 25,000 feet above the surface of the moon. Data and supporting calculations are as follows:

(1) Total time = 1512s + 102s + 151s = 1755 seconds or .49 hours.

(2) What fractional part of an orbit does this represent?

$$\text{Partial orbit} = \frac{.49hr}{1.96hr/orb} = .25 \text{ orbits}$$

(3) Calculate the number of degrees in .25 orbits.

Degrees = (.25)(360°)=90 degrees

You landed in the Sea of Tranquility at 2:53:42 PM.

Moon Chart – Trans-orbit Flight to Surface

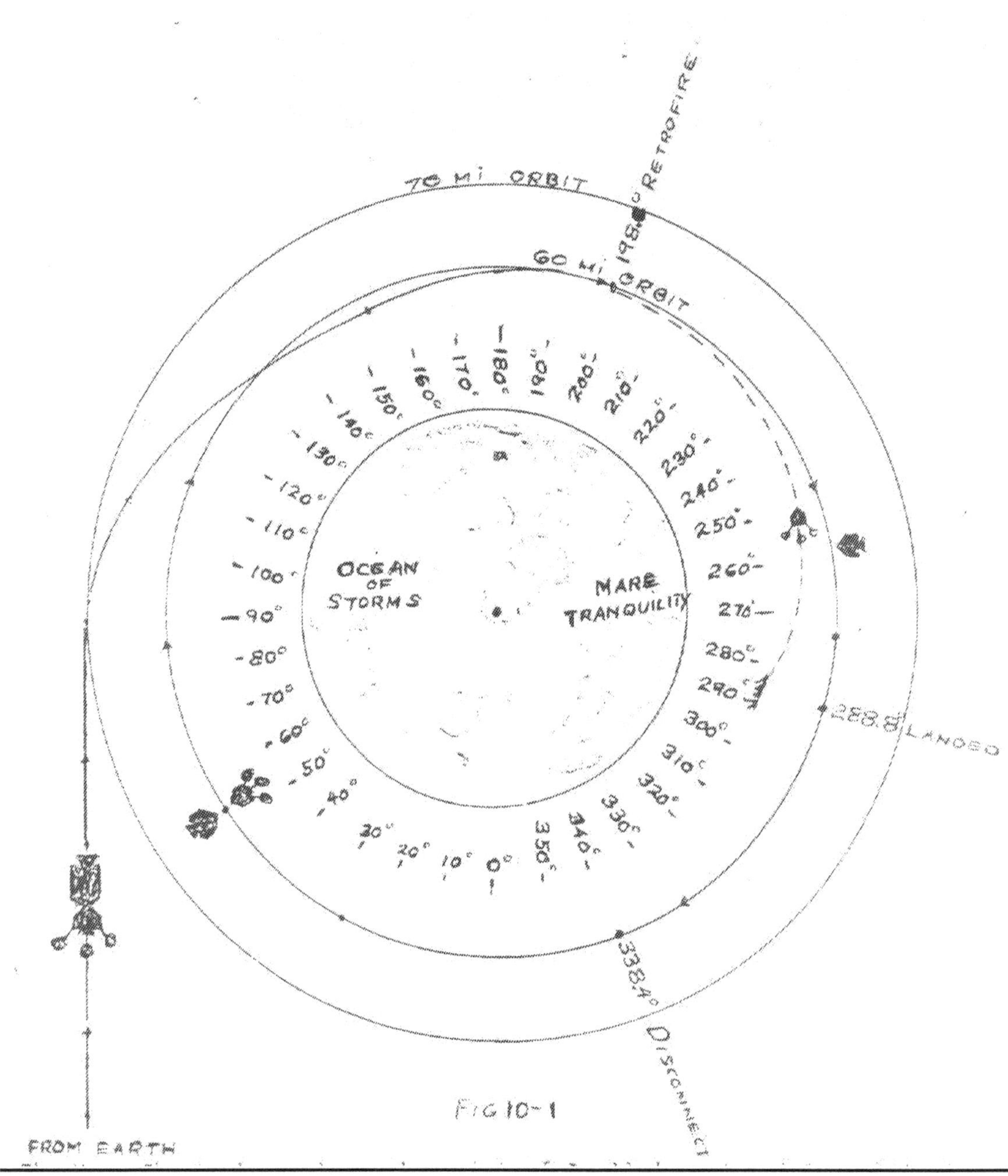

The Moon Chart above shows the point of Apollo's entry into orbit at 70 miles above the moon; the trajectory by which it drops into orbit at 60 miles; the point at which the LM separates from the CM; the point of retro-fire to begin its trip to the surface of the moon and where it lands on the moon. Also, it shows the orientation of each module at any given mode of flight.

Exercise (41)

Self Test

Use the Moon Chart to locate positions of the CM

1. If the CM orbits the moon at 70 miles above its surface at position 90 degrees, and the time is 2:30 PM, answer the following: (a) What is its orbital velocity? (b) What is its period of motion around the moon? (c) If it makes 2.7 more revolutions around the moon, how far will it have gone? (d) What time will it be at the end of this many revolutions? (e) What will be its new angular position?

Slowing Apollo for Descent to the Moon

Apollo is at 24,130 feet above the surface of the moon and moves with a resultant velocity of 1204 ft/sec. Take over the controls. The countdown begins.

```
5 PRINT 2NDBURN
10 LPRITN "AT 25000 FT., RETRO-FIRE FOR FINAL DESCENT"
20 LPRINT "APOLLO-First Manned flight"
30 LPRINT" E   F   N   S   H"
40 S=4.64 'From Page 92, Moon 2B, 1090.64 miles from center of moon
50 T=1
60 E=5282 'Horizontal velocity of LM
70 F=449.7 'From Page 92, Moon 2B, final vertical velocity
80 G=5.28
90 θ=4.87 'Initial angle
100 N=0
110 H=1499 'The horizontal distance, beginning of engine fire
120 LPRINT E,F,N,S,H
130 C=G*1
140 X=E-41*(Cos(θ/57.32))'Horizontal component of velocity
150 B=H+X/5280
160 Y=F+C-41*(SIN(θ/57.32))'Vertical component of velocity
170 D=S-Y/5280
180 G=5.33=(1086/(1086+D))^2
190 θ = ATN(Y/X)*57.32 'Calculation of the new angle
200 N=N+1
210 E=X
220 F=Y
225 H=B
230 IF n=102 GOTO 250 'End the program when time is 102 seconds
240 GOTO 130
250 LPRINT" X   Y   N   D
260 LPRITN X,Y,N,D,B
```

The per-second retardation velocity is based upon the following data and calculations. DATA:

Thrust=30,000p
Impulse=300s
W(ship)=30,000p
W(fuel)=10,645p
V(retro)=41 f/s
Time=102s

CALCULATIONS OF W(fuel)
(V)/(I)(G)=In((W)/(W-Wf))

$$\frac{(102s)(41f/s^2)}{(300s)(32f/s^2)} = \frac{4182f/s}{9600f/s}$$

$$\text{In}\left(\frac{30{,}000p}{30{,}000p - Wf}\right) =$$

Anin=.436=1.55

$$30{,}000\text{p-Wf} = \frac{30{,}000p}{1.55}$$

Wf=30,000p-19355p=
Wf=10,645p of fuel

270 STOP
At 25000 FT., RETRO-FIRE FOR FINAL DESCENT
APOLLO-First manned flight

E	F	N	S	H
5282	449.7	)	4.64	1499
X	Y	N	D	B
1157.889	331.3499	102	4.577245	1560.612

The Final Descent to the Moon

Almost from the beginning, the Eagles landing trajectory was in error as the controlling computer was malfunctioning. Alarms began to sound. Steve Bales, at Houston's Mission-Control, made the decision for Commaneder Armstrong to take control and to attempt a landing. Three was only enough fuel for one attempt

You are now in control of the Eagle and have just passed the landing site at the Sea of Tranquility. The LM is hurling toward the surface of the moon at 3,569 mph. You must quickly orient the ship to an angle of 84.8 degrees with the vertical and fire the engines immediately.

Commander Armstrong was a seasoned pilot with many years of flying experience with all types of airplanes. Hopefully, his spirit rides with you as you fly the Eagle toward a safe landing place.

Aldrin called off the altitude and the velocity of descent as the ship approached the lunar surface:

21,000 feet.............Velocity = 1200 ft/sec
16,000 feet
13,500 feet..............Velocity = 760 ft/sec

At 2000 feet he began to call off the elevation and angle of the ship.

2,000 feet..............47 degrees
1,400 feet..............35 degrees
700 feet..............33 degrees
600 feet................19 degrees

Your landing lights are on and you are frantically searching for a safe place to land the LM. You are surrounded by darkness and the closer you get to the ground, the less you can see of the moonscape because of the great clouds of fine moon dust stirred up by the gas flow form your engines. Now thirty feet above the lunar surface...

...with 30 seconds of fuel left, Armstrong guided the LM between craters and boulders to a perfect landing.[8]

The landing pods are buried a foot deep in the fine moon dust on the surface of the moon. The dust is all around the ship obscuring any vision of the terrain. You look at each other to be sure there have been no injuries. You quickly check to see if the ship is level—the best indication there is no damage. Now you know you are safely on the surface of the moon. A voice from the Command Ship asks if you are down and if everything is OK. You say, "We are safely down, and everything is A-OK." Mission Control is ecstatic and so are you.

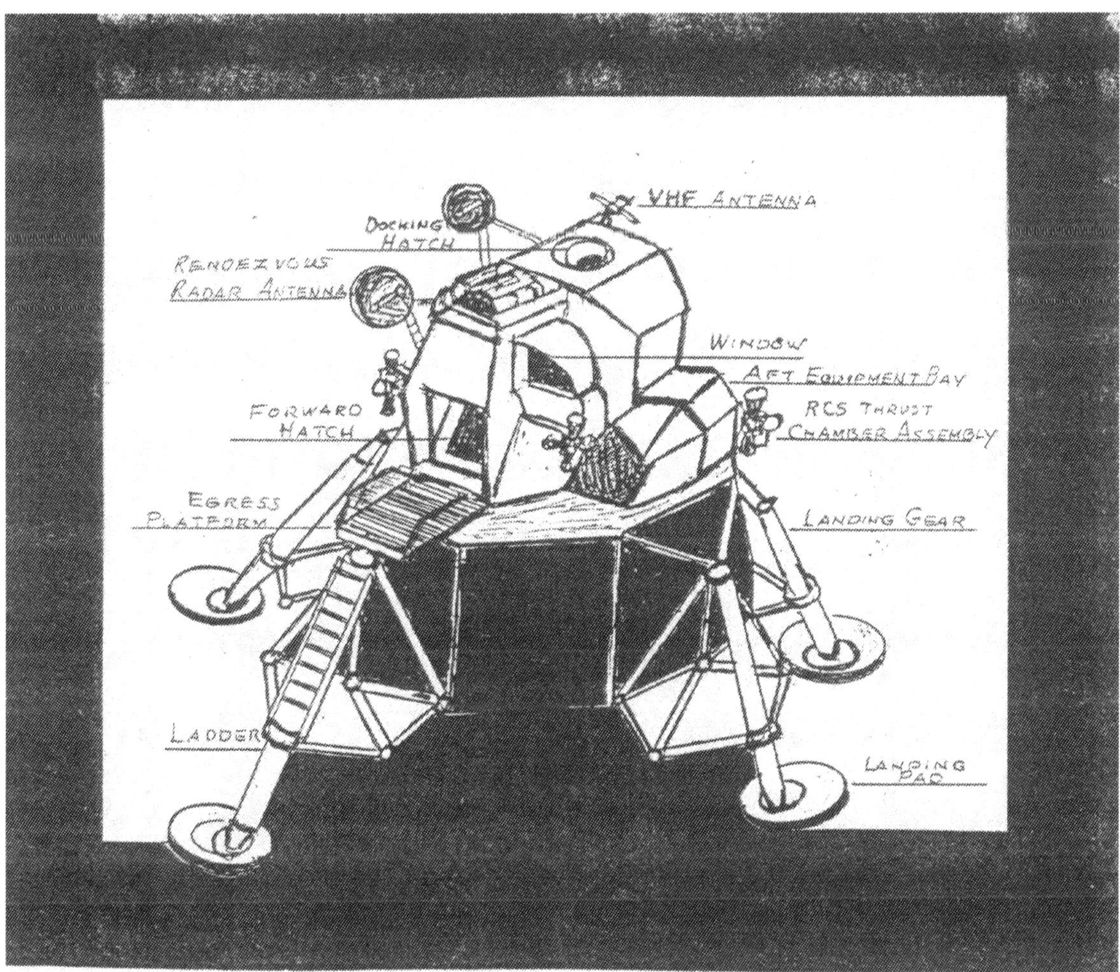

THE 30,000 P-LM AND LANDING BASE

Stands 23 feet high and approximately 11 feet across at its base. Three of the four RCS thrusters can be seen. These thrusters control the vertical and horizontal orientation of the ship as it prepares to land on the moon's surface. The VHF antenna maintains contact with NASA and the orbiting command ship which travels overhead. The rendezvous radar antenna provides a running account of the changing distance to the moon's surface and to the command ship after having rocketed form the lunar surface. The docking hatch at the top makes direct contact with the nose of the command ship and locks at impact. The small craters beneath the ship are typical of millions of others on the surface of the moon.

<u>But Where Did They Land</u>

You had passed over your planned landing site and well beyond. You passed over one very large crater. Lucky you did not land in it—no telling how deep it was. But, where are you? The dust is beginning to clear and the lights pick up a broad pock marked plain. Finally, in the reflected light from earth, you can see the horizon. The dull light makes it difficult to tell how far away it is with any degree of certainty.

> Armstrong: Out of the window is a relatively level plain created with a fairly large number of craters of the five to fifty-foot variety and some ridges, small, twenty or thirty feet high, I would guess, and literally thousands of little one and two-foot craters around…
> Collins to Houston: Do you have any idea where they landed…Just a little bit long. Is that all we know?[9]
> Houston: They may have landed just west of West Crater, BUT THEY WERE NOT SURE! They never were sure where they had landed.

But, where are you? You may never know. You must have passed over Landing Site (2) and apparently you came down in the vicinity of the Crater Schmidt. This location will permit easy access to the Command Module when you return in twenty-two hours.

Houston is detailing what must be done before you leave the LM. You must set all switches for your return to the Columbia. You are advised to be particularly careful about hitting the switches once set, as you may not have sufficient fuel pressure to take off. The problem is your pressure suits. They are so thick and bulky that you can bump into each other in that small space and not even know it.

It is now time to put on and adjust your head-set gear. It will take both of you to do this. Once this is done, you can turn off the oxygen in the module and begin to breathe from your supply an. Once done, exit the ship and be careful going down the ladder. Houston is concerned that you could injure yourself or tear your pressure suit which would be worse. Don't forget, the last step down off of the ladder is three feet. Houston sounds like your mother, but they have good reason to be concerned about your welfare. Bear in mind that gravity is only 1/6th that of earth and that you can jump six times farther and higher than you can jump on earth. The original astronauts were much older than you are, so they were less tempted to high jump and broad jump. You will find your pressure suits cumbersome and moving in them to be difficult. Begin by taking smaller steps and exert yourself as little as possible. If you sweat too much, you will fog up the visor in your helmet and will not be able to see anything.

> Armstrong talking to Houston: "The surface is fine and powdery—I can see footprints of boots." "This is one small step for a man,

THREE FEET TO THE LUNAR SURFACE

Commander Armstrong donned his space suit and helmet with the help of Buzz Aldrin. He then climbed through the hatch of the LM into the moon's darkness and down the ladder. He carefully dropped the last three feet from the bottom of the ladder. Moon dust, only a few inches deep, billowed up around his legs. He left his footprint with every step, knowing they would be there for years to come—to be seen by others who followed. He was in awe of an immense, unbroken silence.

One giant step for mankind."

One Giant Step for Mankind

Aldrin remained on the LM for another twenty minutes monitoring Armstrong's equipment and checking out the astronaut to ship and to astronaut communication system. He then joined Armstrong in setting up three basic experiments having to do with laser beam reflectors, seismic equipment and solar wind measurement. Between them, they collected several pounds of rock specimens to be tested for chemical makeup and age back on earth. Pictures were taken of Aldrin climbing down the ladder and Armstrong setting up the American flag. Finally, they climbed back into the LM and began to make preparations for their ascent to dock with the Command ship in preparation for their trip back to earth.

By this time, the adrenalin has stopped flowing through your body, and you are becoming keenly aware of a deep fatigue that is setting in. You have nearly used up your supply of oxygen and you know there is much to be done before you can blast off to join the mother ship. Although you cannot sit in the Eagle, you can rest against the back of the inner cabin in order to gain some sleep and be ready for the lift off. The last thing you do is to unlock the upper part of the ship from the landers and pods. Once above the lunar surface, you will have no further use for them. That part of the ship which remains for docking is thousands of pounds lighter than the original LM. The final 3500 p thruster is more than enough to return you to the CM. The Basic Program to the left is designed to take your ship from the surface of the moon to with-in eleven linear miles of intersect with the CM. You must pilot the LM to intersect the CM and join with it at orbit, otherwise you will fall back to the surface of the moon.

```
10 PRINT MOONCM
20 LPRINT "Moon Surface to Command
Module"
30 LPRINT "V(Velocity),T(Time),
X(Vx), Y(Vy)"
40 LPRINT"A(Sx),B(Sy),J(Angle),g
(Moon gravity)"
50 G=5.33
60 J=0
70 N=1
80 I=300
90 W=13033
100 M=738*N
110 T=60*N
120 V=INT(9600*LOG(W/(W-M))-
G*60*Cos(J/57.32))
130 X=INT(V*SIN(J/57.32))
140 Y=INT(V*COS(J/57.32))
150 A=INT(A+(X*60/10560))
160 B=INT(B+(Y*60/10560))
170 G=5.33*(1086/(1086+B))^2
180 LPRINT"V="V,"T="T,"X="X,"Y="Y
185 LPRINT "A="A,"B="B,"J="J,"G="G
190 N=N+1
200 IF J>=40 THEN GOTO 230
210 J=(J+10)
220 GOTO 100
230 IF J>=60 THEN GOTO 260
240 J=J+20
250 GOTO 100
260 J=J+10
270 IF J>80 THEN GOTO 290
280 GOTO 100
290 LPRINT "After 480 sec., between
80 and 90 degrees, the ship must"
300 LPRINT"be hand piloted the last
two miles to docking position."
310 STOP
```

Exercise 41

Self Test

The following questions are directed to the Basic program MoonCM on page (110), and to the printout of this program below.

V=239	T=60	X=0	Y=239
A=0	B=1	J=0	G=5.320198
V=839	T=120	X=145	Y=826
A=0	B=5	J=10	G=5.281258
V=1489	T=180	X=509	Y=1399
A=2	B=12	J=20	G=5.214134
V=2194	T=240	X=1096	Y=1900
A=8	B=22	J=30	G=5.120441
V=2960	T=300	X=1901	Y=2268
A=18	B=34	J=40	G=5.011305
V=3834	T=360	X=3319	Y=1918
A=36	B=44	J=60	G=4.923002
V=4744	T=420	X=4457	Y=1624
A=61	B=53	J=70	G=4.845509
V=5741	T=480	X=5653	Y=1000
A=93	B=58	J=80	G=4.803246
After 480 sec., between 80 and 90 degrees, the ship mut be hand piloted the last two miles to docking position.			

(1) What was the weight of the LM at first fire?
(2) By how many seconds is the time incremented? The angle?
(3) What amount of fuel was consumed per firing increment?
(4) What was the final weight of the ship?
(5a) What is the final resultant velocity?
(b) Name the final Vx component.
(c) Name the final Vy component.
(d) What horizontal distance did the ship travel in reaching near orbit?
(e) What was the total vertical distance of the ship?
(6) As used in the program, what does "INT" mean?
(7) Why was the ship not fired again after 80 degrees?
(8) By how much is the final resultant velocity not equal to the orbital velocity?
(9) Copy step (120) of the program and discuss its item by item meaning.
(10) Discuss the meaning of step (150) ie., what does it do?

It is evident that your ship is presently stalled in space two miles form achieving its targeted orbit and moving with a velocity that is 359 ft/sec faster than the orbital velocity of the CM. You must immediately reverse its direction in order to decelerate the craft reduce its speed to that of the on-coming command module. The difference in velocity is quite small, but your LM is moving upward very fast and will pass through the zone of the 60 mile orbit in a matter of a few seconds. You will address this problem on page (137).

LM to CM Three Hours to Rendezvous

It took r480 seconds for the LM to reach the 80 degree transition. At that time, Armstrong took over the controls and began the long, arduous, three hour task of matching the LM to the on-coming CM. These maneuvers had been practiced again and again back on earth. The final maneuvers to join the CM were faultless.

Aldrin: Lift-off from the moon…was exactly on schedule and fairly uneventful. The last two miles before rendezvous we controlled manually, and our intercept

The conversation was getting to be a little bit sober sided. Armstrong: Ascent burn just completed, went quite well, on time at 124 hours and 22 minutes.[11]

Columbia (Collins): Thank you! (The two spacecraft now 91.3 nautical miles apart, closing at 119 feet per second.)[12]

Eagle (Aldrin): Your about 32 seconds later than we are.[13]

Houston: Acquisition of signal, past terminal phase finalization, the two spacecraft are within a few feet of each other, ready to dock.[14]

Eagle (Armstrong): Okay. We're all yours, Columbia.[15]

Columbia (Collins): I'm pumping up cabin pressure. They were docked![16]

THE FINAL ASCENT
Liftoff from the lunar surface was on time and perfect.

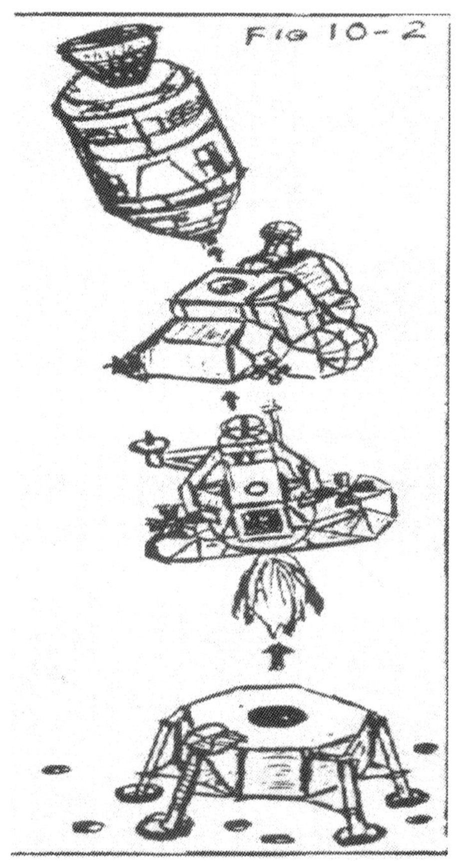

Aldrin: Three hours and ten minutes after lift off we were connected once again with the Columbia. The intervening time had been consumed by long and detailed series of rendezvous procedures and navigational back-up checks…Once we had transferred the rock boxes from the LM voyage back to earth. The LM was sent on its way. It is still orbiting the moon, a reminder of man's first visit.

Looking at the printout of the program from Basic MOONCM on page EX(41) it took the LM 480 seconds to achieve an altitude of 58 miles above the moon's surface. At that time you were moving with a speed of 5,741 ft/sec toward an intersect with the command module. Your ship then reversed direction to continue retrofire in order to slow down to achieve orbit at 60 miles above the moon. It took an additional 10 seconds to reach 59.86 miles above the moon and another 10 seconds to slow to 5,287 ft/sec: Reference the RETRO PROGRAM printout below for the latter information. Your main engine is now stopped and you are in control of the LM. You have made contact with the command module and are moving beneath it, parallel to it. It will take three more hours to dock and lock onto the CM.

On page (142), you will find another method of calculating similar information to that which is included in the Basic program RETRO. This method, which uses quadratic equations, gives a result of 10.94 seconds. The algebraic method does not take the changing mass of the ship into consideration.

The RETRO Basic Program and Printout

As your ship is now in retro (reverse) fire, you would expect the VR velocity to be negative and the acceleration to have become negative or deceleration. Notice that each successive printout of the nine item data is paired and presented in successively paired rows. The cogent items are circled below. Observe that SY is presented in E-03,02 notation after 14 seconds. These items should be zero, but are not because one radian is anot exactly 57.32 degrees.

```
10 PRINT RETRO
20 LPRINT "Use information at the end of MOONCM printout, at 80 degrees."
25 LPRINT "VR,DV,VX,VY,A,SY,SX,T,VORB"
30 W=13033
40 I=300
60 WB=7129
65 J=80
70 FOR T=2 to 20 STEP 2
80 VR=INT(-9600*LOG(W/(WB-DW*T))-4.8*T*COS(J/57.32))
90 DV=INT(5741+VR)
100 VX=INT(5653+DV*COS(J/57.32))
110 VY=INT(1000+DV*SIN(J/57.32))
120 A=INT(DV/T)
130 SY=(2741*T/5280+(.5*A*T^2/5280))*COS(J/57.32)
140 SX=INT((5741*T/5280+(.5*A*T^w/5280))*SIN(J/57.32))
150 LPRINT VR,DV,VS,VY,A,SY,SX,T,VORB
160 IF SY=2 THEN J=90
170 VORB = 5741 +DV
180 IF VORB 5382 Then GOTO 200
190 NEXT T
200 LPRINT "Shut down the main engine."
```

The Last Ten seconds

Make a vector drawing which depicts the position of the LM at 80 degrees. Draw and label the important vectors and angles that have to do with the calculation of the deceleration of the LM to final intersect with the CM and show the distances to the point of intersect.

Data From the Last Part of the Printout of Mooncm
V=5741 f/s; T=480 s: Vx=5653 f/s; Vy=1000 f/s; Sy=58 mi; 0=80°; G=4.8 f/s-s

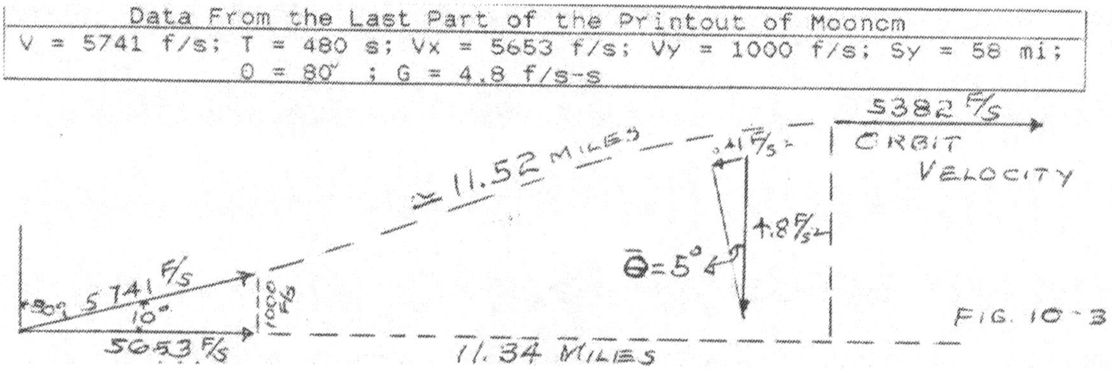

Data From the Last Part of the Printout of Mooncm
V = 5741 f/s; T = 480 s; Vx = 5653 f/s; Vy = 1000 f/s; Sy = 58 mi;
Θ = 80° ; G = 4.8 f/s-s

5382 F/S
ORBIT VELOCITY
≅ 11.52 MILES
.41 F/S²
4.8 F/S²
Θ=5°
5741 F/S
10°
1000 F/S
5653 F/S
71.34 MILES
FIG. 10-3

Next establish two equations that, when solved simultaneously, will result in the calculation of deceleration to the point of intersect with the mother ship and provide the time of travel as the secondary result.

Equation (1) A velocity equation in which (a) is the deceleration and (t) the time to intersect.
Velocity = V – (G)(SINθ)t-(a)(t) = 5382 f/s ⟶ Solve for (t)
Then t = (359 f/s)/(.41 f/s-s + a)
Equation (2) A distance equation using (a) and (t).
Distance = (V)(t) - .5(a)(t)^2 - .5(g)(t)^2 = 60,826 ft
(5741 f/s)(t) - .5(a)(t)^2 = .5(.41)(t)^2 = 60,826 ft (Sub. For t)

$$\frac{(5741)(359)}{(.41+a)}-(a/2)\left(\frac{359}{.41+a}\right)^2-(.205)\left(\frac{359}{.41+a}\right)^2=60{,}826$$

$a^2 - 31.99a = 13.29 = 0$
Use the quadratic formula to solve this equation.

$$a=\frac{31.99\pm\sqrt{(31.99)+(4)(1)(13.29)2}}{}=\frac{31.99\pm\sqrt{1077}}{2}$$

a=-32.4 f/s-s (deceleration); substitute into equation (1)

$$t=\frac{359 f/s}{(.41 f/s-s+a)}=\frac{359 f/s}{(.41 f/s-s+32.4 f/s-s}=10.94\sec onds$$

a=-32.4 f /s-s and t = 10.94 seconds

Exercise 43

Self Test

Solve as directed:

The problem solution on page 142 is based on the last part of the printout shown to the left of Ex. (41). Different, but similar, information is geven in the chart below. For this information and problem, the orbital rendezvous is considered to be at 70 miles above the moon. The chart below gives the pertinent data that you will use to solve the problem written below.

DATA CHART

V = 5515 ft/sec; T = 520 sec; Vx = 5431 ft/sec; Vy = 958 ft/sec; Sy = 67 mi; Sx = 110 mi; θ=80°; G = 4.7 f/s-s; Rmn = 1086 mi.

(1) Make a diagram similar to the one at the top of page (142) and label all parts using the above data as your information source.
(2) Use the equations on page (142) to solve for the deceleration (a) and the time (t).

<u>**Where is the Command Ship at LM Lift-off**</u>

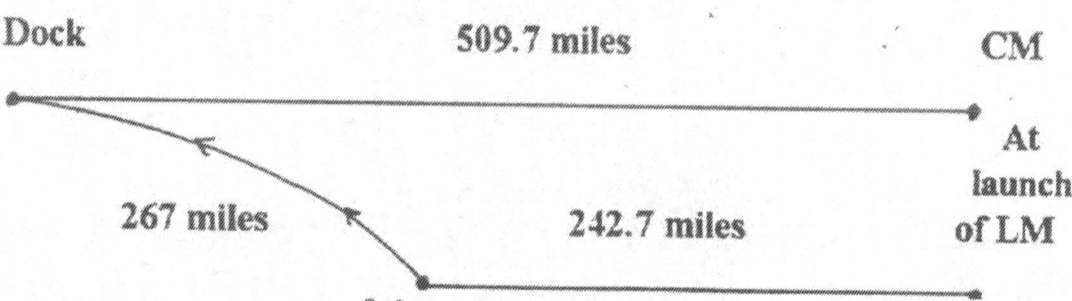

You lifted off from the surface of the moon just 22 hours after the LM began its descent to the surface of the moon. It is now 12:24 PM on July 21st. But, where is the Columbia? Did NASA make the important calculations in order to determine the correctness of the time of departure established by the ship's computer? These calculations are as follows:

(1) Calculate the number of hours from the time the LM began its descent to the moon until blast off from the moon.
Hours = 2:24 PM (7/20) to 12:24 PM (7/21) = 22 hours
(2) Calculate the number of orbits in this period of time.

$$\text{Orbits}=\frac{22 hours}{1.96 hrs/orb}=11.22 orbits$$

(3) Calculate the number of miles in the decimal protion of 11.22 hrs.
Miles = (.22 orbits)(Circum. Of orbit) = (.22 orb)(7197 mi/orb) = 1583 miles

(4) Calculate the number of degrees in 1583 miles.

$$\text{Degrees}=\frac{(1583mi)(360°)}{7197mi/orb}=79.2\deg rees$$

(5) Calculate the longitudinal location of the CM at LM launch.

θ = 198° + 79.2° = 277.2°

(6) If the location of the LM at launch is 288.8 longitude, calculate the longitudinal reading of the intercept point with the CM.

$$\Delta\theta=\frac{(267mi)(360°)}{6{,}820mi}=14.09° \rightarrow \theta = 14.09° + 288.8° = 303°$$

(7) If it takes 500 seconds for the LM to reach orbit, how far will the CM travel in this period of time?

Scm = (Vorb)(T) = (5382 ft/sec)(500 sec) = 509.7 miles

Moon Chart = Lift-off to Orbit Intersect

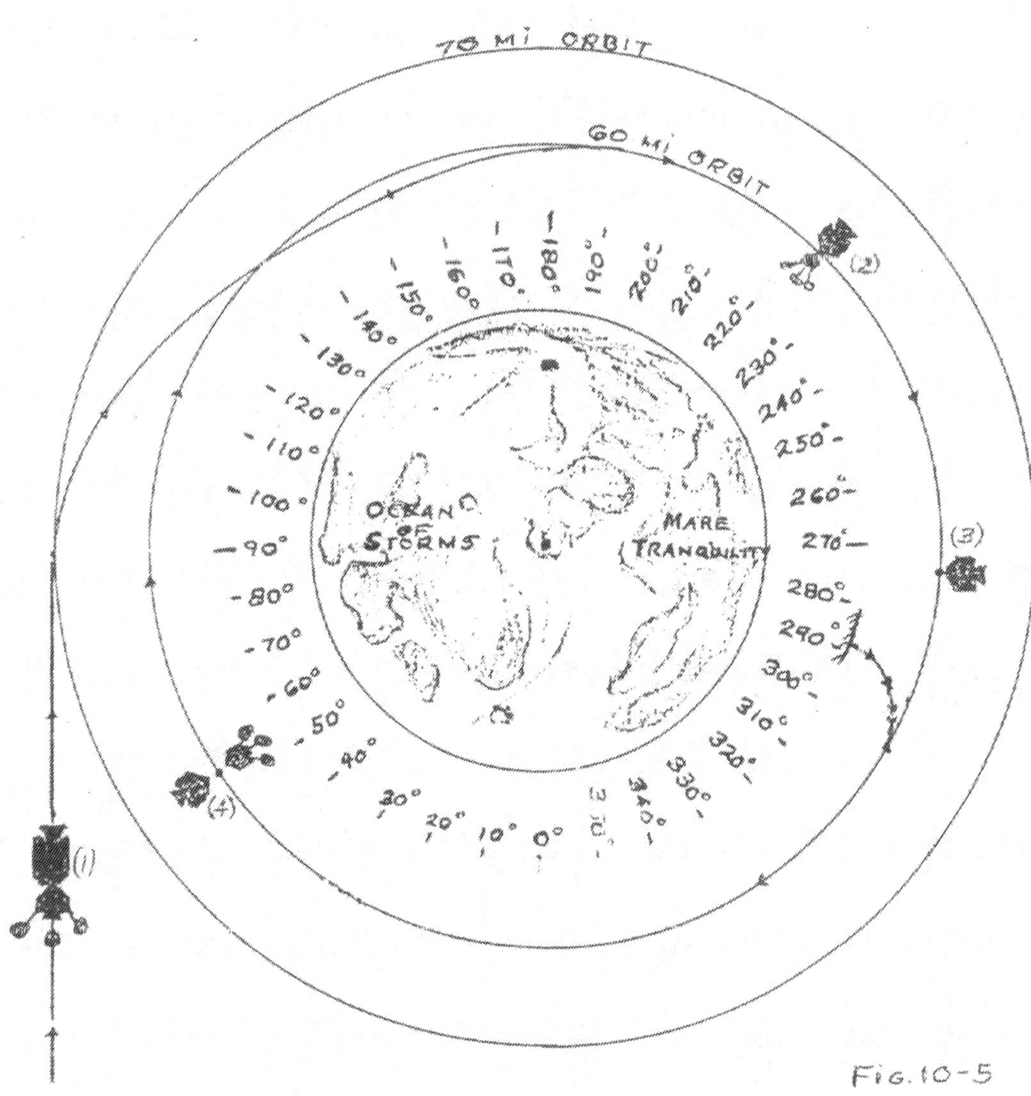

The moon chart above is the same chart as the one on page (132). References are at the moon's equator. All longitudinal designations are at the equator. You are looking at the visible side of the moon. The LM lifted from its surface at 12:24 PM on July 21st, Day VI. The lift off point is at 288.8 degrees on the above chart and (near contact) is not made until at 303 degrees. It took another three hours (1.53 orbits) to complete closure maneuvers and to dock with the command module.

Exercise 44

Self Test

Given the following information to be true about the CM and the LM, solve as directed:

The CM is orbiting at 40 miles above the surface of the moon. The LM disconnected from the CM at 3:30 PM yesterday at 180 degrees and landed on the moon's surface at a longitudinal position of 260 degrees. The LM's main engine fired again to lift it from the moons surface to a position adjacent to the CM in 350 seconds. If the CM was making its ninth orbit, answer the following:

(1) Calculate the orbital velocity of the CM in ft/sec and mi/hr.
(2) Calculate the circumference of the orbital distance in miles.
(3) Calculate the orbital time of the CM.
If the CM is in its ninth orbit and if the LM travels a horizontal distance of 210 miles in order to connect with the command vessel, answer the following:
(4) Calculate the average resultant velocity of the LM.
(5a) What is the fractional part of the CM's orbit at the time of LM lift off? (b) What is its longitudinal designation at this time? (c) What was the designated time at lift-off?
(6) How far away was the CM from an orbital point directly above the position of lift-off?
(7) What was the longitudinal position of both crafts when they first achieved parallel position?
(8) If it took an additional 1.5 hours to achieve a locking position, (a) What is their new longitudinal position? (b) How many linear miles did they travel? (c) What time is it?
(9) Make and designate angular positions on a chart similar to that on page (144)B.
(10) Make drawing similar to that on page (144) and label all appropriate designations.

Together Again

The time is 12:32:20 PM on July 21, 1969. The command ship hovers above you. NASA is ecstatic, you are a few feet short of joining with the CM. After hours of precision maneuvering, the two vessels are now locked together. Upon locking, the three of you warmly shake hands and embrace, relieved that you are safely back together again. But time is short-you quickly transfer precious rocks and equipment to the main vessel, close the hatch to the LM, and with the push of a button detach it from the CM to forever circle the moon.

A Basic program to Escape Moon's Gravity

If you are to get home, you must write a program that your computers can use to get you there. It is necessary that you write a correct "Escape from the Moon" Basic program. In order to do this, there is much data to be assimilated. These are the important facts:

Moon's Radius = 1086 mi	Escape Velocity = 7818 ft/s
Orbital Distance = 60 mi	Moon's Gravity = 5.33 ft/s-s
Orbital Velocity = 5382 ft/sec	Ship's Weight = 40,841 p
Fuel Weight = 20,000 p	Engine Thrust = 22,000 p
Impulse = 410 sec	Moon's Velocity = 2162 mi/hr

Additional information that must be calculated is as follows:

(1) Time of engine burn (T)	(2) Final velocity of the CM (VR)
(3) Longitudinal angle of launch	(4) Distance from point of launch (S)
(5) Distance form center of moon (C)	(6) The angle (AN)

(The angle An is the angle between the perpendicular to ship's launch point and the distance from the ship to the center of the moon at any point during travel.)

You already know the basic equations that you will need to solve this problem and to calculate the above information. These equations are as follows:

VR = (Vex)(ln R)) – (G)(T)(Cosθ) and Impulse = (F)(T)/(Wf)

There are too many variables in the above equation to make for a simple Basic program solution to this problem. With one less variable, you can then write a reasonably accurate program to do this. It is possible to reduce the fuel weight to a function of (T) by solving the impulse equation for (T).

$$T = \frac{(I)(Wf)}{(F)} = \frac{(410s)(20{,}000p)}{22{,}000p} = 373\,\text{sec}onds$$

This tells you that it took 373 seconds to burn 20,000 p of fuel. Therefore, the engine burns 54 pounds of fuel a second. Now Wf = 54 pounds per second and R becomes (W) / (W-54 (T)).

(Continue this discussion on pg 148.)

Exercise 45

Self Test

Show all calculations:

Your ship is in orbit at 70 miles above the moon and is at the extreme perigee. It will be your advantage to refer to page (112) to figure (8-4).

(1) What is the velocity of the moon in mi/hr and ft/sec at the above location in its orbit?

(2a) Calculate the escape velocity form the moon in ft/sec and mi/hr at this location. (c) Is this different from the escape velocity at the apogee?

(3a) Calculate the orbital velocity of your ship. (b) What additional velocity will your ship need in order to achieve escape velocity from the moon?

Make use of the following data in performing the remaining problem exercises: I = 400 s; Wt = 41,000 p; G = 4.7 f/s-s; θ = 86°; F = 22,000 p.

---

(4) Calculate the amount of fuel (Wf) needed to make this burn. Use the equation on page (92) to make this calculation. (Remember that this equation is for free space and will provide only an approximate answer where gravity is involved.)

(5) Calculate the time of burn (T).

(6) Calculate the final velocity of your ship at last burn.

(7) By how much does your calculation differ from the correct escape velocity? (Remember that under actual conditions, you would have ample fuel in reserve to make up for the difference.)

(8) What was your acceleration during last burn?

(9) How far from the surface of the moon is your ship at burnout?

(10a) Use an average gravity of G=1.81f/s and calculate how much farther the ship will go into space due to its final velocity at burnout. (b) Calculate the retardation distance imparted by gravity.

$$R = \frac{W}{((W - 54(T))}$$; now VR = V(2) + 5382 where 5,382 ft/s is the orbital velocity of the ship.

The VR equals:

VR = 5,382 + (G)(I)Log((W)/((W) – 54 (T))) – (g)(T)Cosθ

This equation can be simplified by letting DRAG = (g)(T)Cosθ and by letting (G)(I) = (32 f/s-s)(410 s) = 13,120 f/s. The final equation is much simpler.

$$VR = 5382 \text{ f/s} + (13{,}120 \text{ f/s}) \text{ Log}\left(\frac{W}{W - 54(T)}\right) \text{-DRAG}$$

It is not necessary that you understand all of the facets of the development of this equation. It is important that you be able to understand how it is used in a BASIC program. There is one other topic to be investigated prior to firing the rocket.

Reaching Escape Velocity from Orbit

The drawing to the right shows the direction of the CM being fired from orbit. Notice that its direction is tangent to the orbital line of flight and that it is perpendicular to a line drawn from the center of the moon to the point of ignition. Observe that the moon is moving away from the line of fire of the rocket at 2162 mi/hr. The line AC, which completes the right triangle ABC is always the distance from the CM to the center of the moon and is used in the program to determine the new gravity at the (C) position. The angle ABC is the angle used in the DRAG FORMULA in the equation.

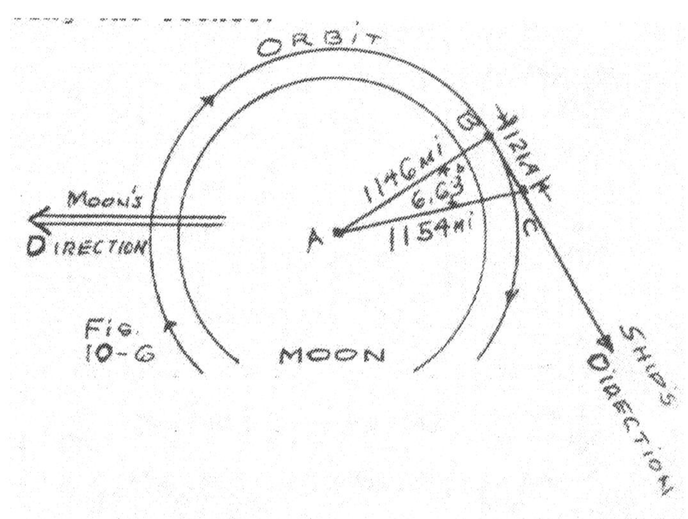

A Basic Program to Achieve Escape Velocity

```
10 LPRINT "T        VR      S       C           AN"
20 FOR T = 0 TO 200 STEP 10
30 W = 40841! 'W is the weight of the rocket
40 V2 = 5382 + 13120 * LOG(W/(W –54 * T) 'Final velocity without DRAG
50 D = V2 * T/10560 'This is twice the number of feet in a miles
60 A = ATN(D/1146) 'This is the angle whose tangent is this quotient
70 C = 1146/COS(A) 'This is AC in the drawing
80 G = 5.33 * (1086/C)^2 'This is the new positional gravity of the ship
90 DRAG = G * T * COS (A)
100 VR = V2 = DRAG 'This is the final corrected velocity
110 S = VR * T/15060 'This the distance BC in the drawing
120 AN = A * 57.3 'This changes the angle from radians to degrees
130 LPRINT T, VR, S, C, AN
```

140 IF VR > 7818 GOTO 160 'This is the conditional statement
150 NEXT
160 STOP
Printout from the Escape Velocity Program

T	VR	S	C	AN
0	5382	0	1146	0
10	5508.765	5.216634	1146.012	.2630961
20	5637.898	10.67784	1146.051	.5429396
30	5769.471	16.39054	1146.123	.8398573
40	5903.57	22.36201	1146.232	1.154183
50	6040.28	28.59981	1146.386	1.486258
60	6179.697	35.1119	1146.589	1.836429
70	6321.92	41.90667	1146.849	2.205048
80	6467.058	48.99287	1147.174	1.592473
90	6615.227	56.37977	1147.571	2.999067
100	6766.552	64.07719	1148.06	3.425195
110	6921.164	72.09545	1148.621	3.871225
120	7079.208	80.44553	1149.291	4.337529
130	7240.839	89.13911	1150.074	4.824477
140	7406.212	98.18841	1150.98	5.332433
150	7575.513	107.6067	1152.023	5.861771
160	7748.923	117.4079	1153.215	6.412846
170	7926.646	127.607	1154.57	6.986021

DATA RESULTS

From the end of the data above, you can tell that the program did not complete the loop, but separated from the loop between 160 s and 170 seconds. By a process of interpolation, the escape velocity of 7818 ft/s was reached in 163.9 s and the distance traveled at that time was 121.4 miles. At that time, you were 1153.7 miles from the center of the moon or 67.7 miles from its surface. Look at the moon drawing on Page 148 to have a picture concept of this part of your return to earth maneuver. You reached 7818 ft/s so quickly because the moon's gravity is only 1/6 that of earth and because your orbital velocity was 5382 ft/s.

The Optimum Firing Point – Moon to earth

If you were attempting to cross a wide river flowing north to south and you wanted to row from Point A to Point B on the other side, you must begin at a point well above A in order to reach B. Otherwise the current will take you below your destination.

On the day of departure from the moon, the moon's velocity is 2,162 mi/hr. When NASA fires the rocket engine to escape the moon's field of gravity, the rocket will be propelled in the direction in which the moon is traveling, at that velocity. Just as your boat would have missed your landing point, so would Apollo miss earth if fired directly at earth. It is necessary to calculate the magnitude of this distance and to incorporate it into the planned trajectory of your ship.

From Escape Velocity to Gravity Zero

You are on your way back to earth after having achieved escape velocity from the moon. Your longitudinal point of fire from moon orbit was 45.8 degrees. This point of departure from orbit takes into account the deep space drift of your ship due to the moon's velocity of 2162 mph as it circles earth each month.

The Basic program below is designed to monitor your ship's progress after the completion of its acceleration to escape velocity, until it reaches zero gravity at 30,000 miles. Upon reaching zero gravity, earth gravity will then take over. At that time the attraction of the earth for your ship will equal that of the moon. You will pass through the zero gravitational interface moving with a velocity of 3289.4 ft/s. Your ship will then begin to accelerate toward earth until it reaches a velocity that is approximately that of escape velocity from earth. The BASIC program below will provide this information for you. Run this program and observe that it takes about 40 seconds to process.

```
10 LPRINT "MOONTO 0 – to ZERO GRAVITY INTERFACE"
20 LPRINT "APOLLO – Back to Zero Gravity"
30 LPRINT "VELOCITY   DISTANCE   TIME    GRAVITY"
40 A = 7818 'Your ship has reached escape velocity (Data Page 119)
50 S = 121.4 'This is your distance out from point of ignition
60 G = 4.72 'Moon gravity is 4.72 ft/s-s
70 V = 7818
80 P = 0
90 T = 0
100 LPRINT V,S,T,G
110 C = G*1
120 V = A – C
130 D = (1*V/5280) + P + S
140 R = 1146 + D
150 G = 5.33 * (1086/R)^2
160 T = T + 1
170 P = D
180 S = 0
190 A = V
200 IF D > 30000 THEN GOTO 220 'You are now 30,000 miles from the moon
210 GOTO 110
220 LPRINT V,D,T,G
230 STOP
MOONTO - = ZERO GRAVITY INTERFACE
APOLLO- Back to Zero Gravity
```

VELOCITY	DISTANCE	TIME	GRAVITY
7818	121.4	0	4.72
3289.359	30000.61	41792	6.479855e-03

Chapter XI

Return to Earth

In Retrospect – The Optimum Direction of Fire

The Apollo CM and SM are in free space traveling (toward earth??). You are moving with a velocity of 3289 f/s away from the moon, but your drift is parallel to the moon and has a magnitude of 2162 mi/hr. As with the example of the boat crossing a river, your direction of launch from the moon must take into consideration the quantity of moon drift imparted to your ship. Using the results from the printout of the RTNFRESP program below, in reference to the drawing to the right, the CM and SM drifted 145,935 miles while traveling 211,000 miles to earth. This took 67.5 hours after fire from moon orbit. This makes the angle theta equal to 34.24 degrees, and, makes your orbital point of engine ignition 145.8 degrees from your first entry into orbit around the moon.

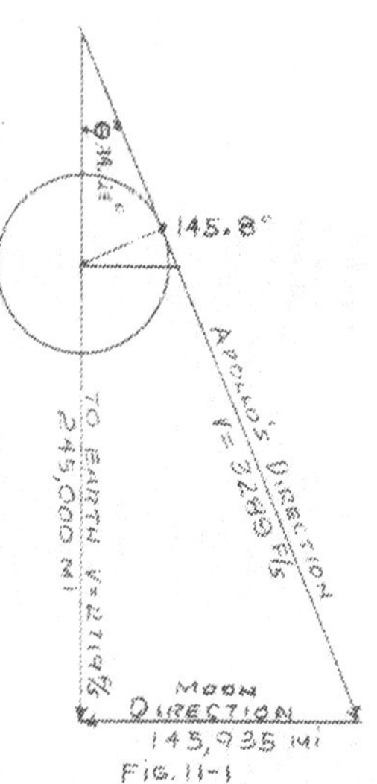

FIG. 11-1

A Basic Program to Return Apollo to Earth

```
10 LPRINT "RTNFRESP – FROM THE ZERO GRAVITY INTERFACE TO EARTH"
20 LPRINT "APOLLO – First Manned Flight to the Moon"
30 LPRINT "VELOCITY   DISTANCE   TIME   GRAVITY"
40 A = 2719 'CM component of velocity at gravity interface
50 T = 0
60 S = 211000! 'Distance to the surface of the earth
70 G = .011 'Earth gravity at the interface
80 P = 0
85 V – 2719 'Velocity component
90 LPRINT V,S,T,G
100 C = 1*G 'Gravity converted to velocity
110 V = A + C
130 P = (1 * (V/5280)) 'Distance per second of time
140 D = S – P
150 S = D
160 G = 32 * (4000/D)^2 'The new gravity
170 T = T _ 1 'Intervals are one second
180 IF D < 4300 THEN GOTO 200 'End the program at 300 miles above earth
190 GOTO 100
200 LPRINT V,D,T,G
210 STOP
```

RTNFRESP – FROM THE ZERO GRAVITY INTERFACE TO EARTH
APOLLO – First manned Flight to the Moon

VELOCITY	DISTANCE	TIME	GRAVITY
2719	211000	0	.011
35198.02	4297.085	201244	27.72823

Exercise 46

Self Test

Continue to reference page (113) for supportive information to this exercise. Otherwise, complete problems and questions as directed.

Part I Reference BASIC Program "Escape Velocity"

(1) Is the extreme apogee located at 245,000 miles from the center of earth?
(2) How far from earth is the extreme apogee?
(3) How fast is the moon moving away from the direction of rocket fire?
(4) How many miles does the rocket travel during burn?
(5) What is the time of burn?
(6) How far is the rocket form the center of the moon at burnout?
(7) What is the change in velocity during burn?

Part II Reference BASIC Program "Moon to 0"

(8) At what distance is zero gravity from the moon's center?
(9) What is your velocity at this point?
(10) How many hours does it take the CM to arrive at this location?
(11) Why is 30,000 miles form the center of the moon considered to be zero gravity at this location?

Next to Last Schedule for Landing in the Pacific Ocean

You have finally left the moon-earth, zero gravity, interface. You have traveled in free space, under the pull of earth's gravity, for 55.9 hours. At this moment, you are 297 miles from the surface of a gigantic earth and your ship is moving at 35,198 ft/sec. What time is it and what is the date? Apollo 11 landed in the Pacific Ocean on July 24, 1969, at 11:22 EDT. Is your ship on or off schedule?

In order to re-establish your schedule, let us return to July 20th when the Lm disconnected from the CM at 12:26:42 PM, after which it landed on the moon at 2:53:42 PM. The following day, the LM blasted away from the moon's surface. The schedule is as follows:

Day VI – July 21/69 at 12:24 PM (Blast off from the moon's surface at 0.139 hours to a 60 mile orbit. Time at 12:32:20 PM.	(Continued from below) Its longitudinal position was 145.8 Degrees and the time was 3:36:10 PM.
Two hours to docking-the time Was 2:32:20 PM at 1.02 orbits, At 310.3 degrees.	Day VII-11.65 hours later, the ship Reached zero gravity at 3:15:10 AM
After docking, the Apollo made 0.543 orbits in 1.064 hours. At that time it was in position to fire its engines to achieve Escape velocity	Day VII and Day VIII, the ship was In free space arriving at 297 miles From earth in 55.9 hours at 11:09:10 AM. July 24, 1969 was the day that Apollo 11 landed in the Pacific Ocean.

Your ship is running very, close to being on the original schedule. Considering the length of the trip, NASA and your crew have done very well. The following calculations will support the above schedule and the previous BASIC program.

Calculate the time in hours from moon orbit to earth entry.

Total time = Moon T = Earth T = $\dfrac{41{,}940s + 210{,}244s}{3600s/hr}$ =11.65 hr + 55.9 hr = 67.55 hr

Calculate the ship's drift distance imparted by the moon.

D = (Vm)(T)=(2162 mi/hr)(67.55 hr) = 146,043 miles

Calculate the angle of moon launch and the angle of departure.

$\cos\theta = \dfrac{Vmn}{Ver} = \dfrac{2719 f/s}{3289 f/s}$ ⟶ θ = 34.24 deg and θ (dep) = 180°-34.24° = 145.8 deg.

Exercise 47

Self Test

Assume that your ship is returning to earth from the vicinity of the extreme perigee of the moon at orbit 60 miles. The data will be similar to that of the apogee except for the moon's distance form earth.

(1) Make a similar drawing to that on P-151, but do not label its parts until you finish this exercise.
(2) At the perigee, how far is the moon from earth?
(3) What is the moon's velocity at its perigee?
(4) Use a simple ratio to solve for gravity interface distance at the perigee.
(5) Given time to zero gravity is 11.61 hours at 30,000 miles form the moon, how far would you expect it to be at the perigee?
(6) If the time from zero gravity to earth is 55.9 hours: (a) Calculate the distance of moon drift during your trip to earth from the moon. (b) How many days back to the proximity of earth? (c) What drift would you expect your ship to under-go during this same period of time?
(7) What total distance does the CM travel during this period of time?
(8) Calculate the angle theta in your drawing.
(9) Now label all parts of your drawing.
(10) At what angle in your orbit about the moon did you fire the CM to leave moon orbit?

A Basic Program to Safely Bring the CM to with-in 6.6 miles of Earth

The Basic Program below provides an algorithmic analogue of your descent from 297 miles above earth to within 6.6 miles of the Pacific Ocean. The CM, moving with a velocity of 35, 198 ft/sec, has its velocity reduced to 15,547 ft/sec in 337 seconds. This is an average deceleration of 58.31 ft/s-s. This amount of deceleration is easily withstood by human beings, so you should not have felt excessive discomfort during this part of the trip. The CM traveled a horizontal distance of 904 miles while it dropped a vertical distance of 290.35 miles.

In this program, F, K, and M represent frictional deceleration factors caused by gaseous particles in the exosphere. The (F) factor acts upon the horizontal component of the ship's velocity. The (K) and (M) factors act upon the vertical component of velocity which is increasingly effected by gravity. You will recall that the greater the velocity of an object, the greater the drag force upon it. The CM has a very great velocity upon entry, but gaseous space is very thin. These two factors compensate to bring about a quick but gradual reduction of velocity. The third factor (M) is introduced after the CM has dropped 104 miles, because the gaseous particles are becoming more dense and , therefore, gaseous friction is greater.

```
1 LPRINT "APOLLO 11, the Final Descent – DXDY"
2 LPRINT "This begins the entry sequence of the Command Module from"
3 LPRINT  "space.  The CM enters earth space at 6 degrees below a"
4 LPRINT "radial normal at 297 miles above earth.  It will drop to"
5 LPRINT "within 6.6 miles of the ocean in 337 seconds."
10 DATA 0,0,0,0,0,0,3679,35005,27.63
20 READ DY,DX,F,T,K,M,V,R,J
30 F = F + .35 'The horizontal frictional deceleration factor.
40 T = T + 1
50 K = K + 27.7 'The vertical frictional deceleration factor.
60 VX = R – F 'The horizontal distance equation.
90 VY = V + J – K – M 'The vertical velocity equation.
100 V = VY
110 DY = DY + VY / 10560 'The vertical distance equation.
120 G = 32 * (4000 / (4297 – DY))^2 'Gravity calculation.
125 J = J + G
127 IF T>=240 GOTO 160 'Time decision statement.
130 IF T = 200 GOTO 145
140 GOTO 30
145 LPRINT "T   VX   VY   DX   DY"
150 LPRINT T,VX,VY,DX,DY
160 M = M + 9.2 'Second vertical deceleration adder for more dense gas.
165 IF DY >=290 GOTO 180 'Final inequality distance statement.
170 GOTO 30
180 LPRINT T,VX,VY,DX,DY
190 STOP
```

Printout from DXDY Basic Program

Apollo 11, the Final Descent – DXDY

This begins the entry sequence of the Command Module from space. The CM enters earth space at 6 degrees below a radial normal at 297 miles above earth. It will drop to within 6.6 miles of the ocean in 337 seconds.

T	VX	VY	DX	DY
200	27970	10943.51	618.1168	103.9443
37	15071.45	3817.987	903.8064	290.3515

The Final Basic Program to Splashdown

This program will take you to within 185 feet of the water. Your velocity at this elevation will be 55.7 ft/sec or 37.9 mi/hr. You could safely hit the water at this speed, but the chutes will slow down the CM to 20 mi/hr at splashdown. As this program is similar to the latter, it is listed below without further explanation.

```
5 LPRINT “Apollo 11, The Final Descent – XDX”
10 LPRINT “This program will take the Command Module from 6.6 miles”
20 LPRINT “above earth to within .035 miles of its surface.  Drogues”
30 LPRINT “popped out at 5.5 miles above earth to slow your descent.”
40 LPRINT “Parachutes will come out about 1 mile above earth.”
50 DATA 125, 337, 31.89, 30.1, 6.6, 904, 15071, 3818
55 READ F,T,J,K,DY,DX,R,V
60 F = F + 26.83
70 K = K + 39.15
80 T = T + 1
90 VX – R – F
100 R = VX
110 VY = V + J – K + M
120 DX = DX + VX/10560
170 V = VY
180 DY = DY – VY/10560
190 G = 32 * (4000/(4006.6 – DY))^2
200 J = J + G
210 IF DY <= .41 GOTO 225
220 GOTO 60
230 LPRINT T,VX,VY,DX,DY
240 LPRINT “PARACHUTES ARE OUT”
250 M = +263
260 IF T = 378 GOTO 280
270 GOTO 70
280 LPRINT T,,VY,,DY
282 LPRINT “At 378 seconds, the CM has a velocity of 37.9 mi/hr and is”
284 LPRINT “185 feet above the ocean.  You will splash down about now!”
290 LPRINT “SPLASH DOWN”
300 STOP
```

Printout from the XDX Basic Program

Apollo 11, The Final Descent – XDX

This program will take the Command Module from 6.6 miles above earth to within .035 miles of its surface. Drogues popped out at 5.5 miles above earth to slow yoru descent. Parachutes will come out about 1 mile above earth.

T	VX	VY	DX	DY
364	1554.256	283.3352	928.7756	.4051343
PARACHUTES ARE OUT				
378		55.65235		.0349638

At 378 seconds, the CM has a velocity of 37.9 mi/hr and is 185 feet above the ocean. You will splash down about now! SPLASH DOWN

The Last 185 Feet

It would have been fairly easy to append the last program to bring the CM the rest of the way to the ocean, but later on, you will be asked to do that as an additional exercise. Although 61.67 yards is about 2/3 the length of a football field, or about the height of a nine story building, I doubt you would care to jump into water form that elevation. But, suppose you did jump from that height. How long would it take for you to impact the water? That should be an easy calculation for you. How about 3.4 seconds. With what final velocity would you impact the water? Your final velocity would be 108 ft/sec, or about twice the velocity of the CM at 185 feet above the water. Of course, your calculation did not take into consideration that you were already moving at 55.65 ft/sec, or that you had three large parachutes attached to the back of your anatomy.

The printout above has important information about your deceleration during the last 14 seconds of descent. The vertical component of velocity changed from 283.33 ft/sec to 55.65 ft/sec. Calculate the average deceleration of the CM.

Decel = (V2 – V1)/T = (283.33 ft/s – 55.65 ft/s)/14s = 16.26 ft/s-s

The average deceleration is 16.26 ft/s-s. The CM will reach terminal velocity at 20 mi/hr or 29.33 ft/sec. It will descend the remaining distance at that velocity. In how many seconds will the CM reach terminal velocity?

T = (V2 - V1)/A = (55.65 ft/s – 29.33 ft/s)/16.26 ft/s-s = 1.46 s

How far will it have fallen in this length of time?

$S = .5(A)(T)^2+(V)(T) = -.5(16.26 \text{ ft/s-s})(1.46s)^2+(55.65 \text{ ft/s-s})(1.46f/s)$

S = 69.87 ft

The remaining distance to fall at 20 mi/hr is 115.13 ft. You make the final calculation of time for this distance.

You must have wondered about the horizontal velocity and distance. Note that VX is 1554 ft/sec. Remember that the velocity of a point on earth is 1535 ft/sec. The VX velocity is close enough to this number that the CM is in vertical drop and has no horizontal component with regard to rotating earth.

Relative to Your Departure from Earth – Where Are You

Geographically, where are you on earth? You could have landed in China. Since you landed in water, you must be somewhere in the Pacific Ocean, or did you land in water? The point at which you enter earth space determines where you will land on the surface of earth. We will assume that your entry is in the Northern Hemisphere along the line that represents 10° latitude on a world map. This will place the CM in the locale of Hawaii and the Johnson Islands

Not long ago, you did considerable work with a moon chart using it to keep track of the LM and the CM as they circled the moon. On the next page is an earth chart (center). This chart is marked off in units of ten degrees and rotates counter clockwise. The larger, outer chart is a fixed chart, similarly marked, but with (0) degrees pointing directly east. When both zeros set opposite each other, there is a direct one to one correspondence, and it is precisely at this moment that it is midnight in Greenwich, England. All world time is derived form this measurement. This is true, because the primary earth meridian (0 degrees longitude) passes through Greenwich. The fixed chart makes it easier to keep this fact in mind. Because the time reference in Greenwich is Standard Time, all remaining references to time will be Standard rather than DLS time.

Looking at the chart, you will see that Kennedy Space Center (KSC) is located at Longitude 81 degrees. The designation at the bottom of the chart is the location of KSC when it is midnight in Greenwich, England. By the time Apollo 11 was launched, earth had rotated to the second position (upper left), which is 226 degrees. By 8:32 Am, earth had rotated 215 degrees from the lower location, or to a position on the opposite side of the earth. Remember that each angular designation represents a complete circle that moves in and out of the paper. All circles join at a given point above and below the paper. These points make up the North and South Poles of earth.

The easiest way to locate your point of entry into earth space is to determine the time at which you arrived at the top of the circle, and then to calculate how much farther earth had rotated by your time of arrival. Your return to earth was only a few minutes earlier than the actual crew of Apollo 11. For that reason, your landing point will be near that of the original splashdown. The difference between 8:32 AM (your time of departure from earth) and 11:09 AM(your time of return) is 2 hours and 37 minutes. In the time you were gone from earth, it had rotated eight complete revolutions plus an additional distance due to your later arrival. This distance amounts to 39.3 degrees. KSC is now located at 186.7 degrees on the fixed map. As you entered space at the top of the circle, your point of entry was at 125 degrees less 6 degrees plus 39.3 degrees longitude on the rotating circle. The 6 degrees takes into consideration that the ship entering earth space at an angle of 6 degrees. Where the lines representing 158.3 degrees longitude and 10 degrees latitude cross on a world map, designates your point of entry into earth space. This position is between Hawaii and the Johnston Islands, in the Pacific Ocean. Added to this distance is an additional 818 miles that the CM traveled due to horizontal displacement during this drop to the ocean. This amounts to an additional 11.7 degrees of displacement on the map. Why isn't this distance 929 miles?

Earth Longitudinal and Angular Charts

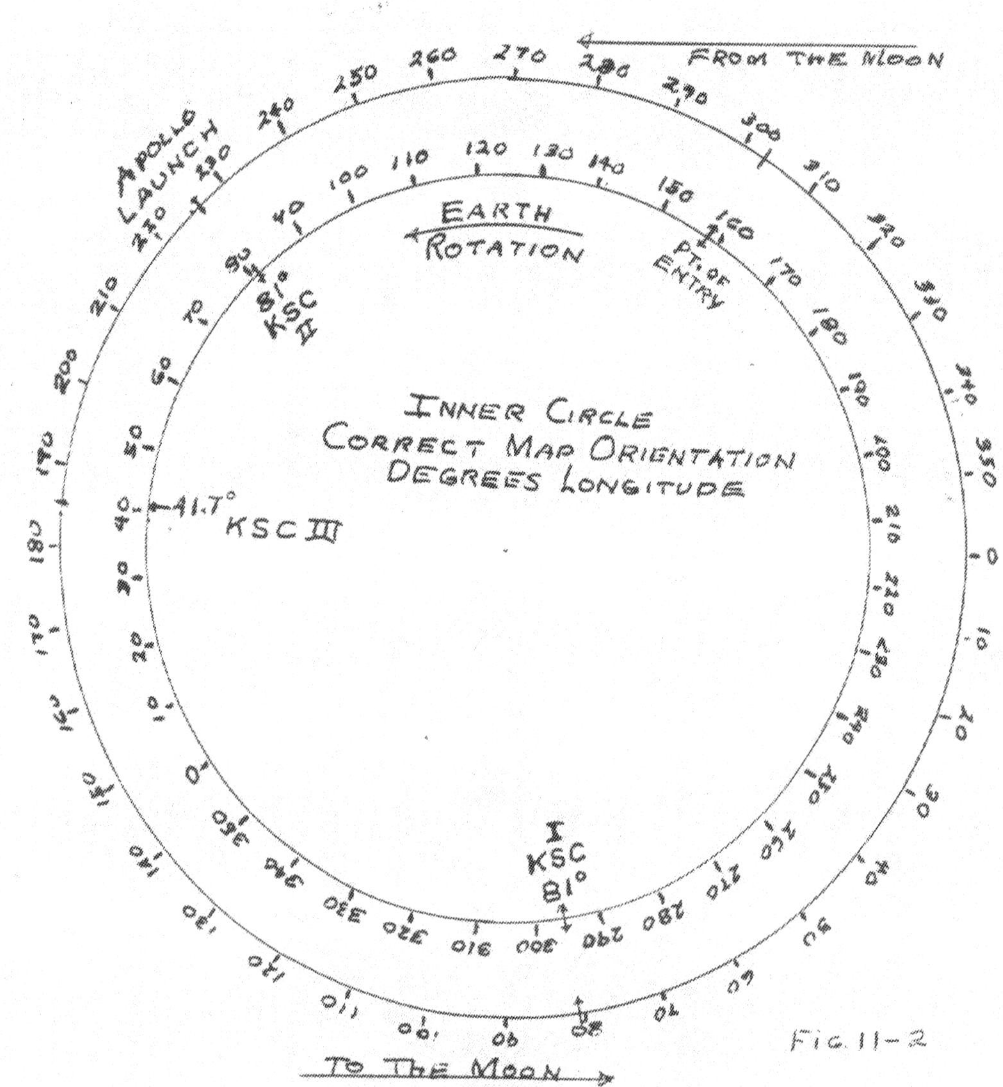

Exercise 48

Self Test

This is a fundamental exercise based upon the fixed and rotating charts on the opposite page. Use the charts to reference your work when necessary. Some of these problems will be a review of circular motion.

1a. In which direction does the earth turn? b. What is the direction of the rotating chart?

2. When the fixed chart and the rotating chart are set like angle to like angle, what geographic location is at θ degrees?

3. Assuming both charts are set at zero degrees and then the earth chart begins to move from (a) to (b) to (c) locations, what will be the readings on the fixed chart under the following conditions: (a) The earth chart moves counter clockwise 55 degrees. (b) The earth chart now moves 105 degrees in the opposite direction. (c) It now moves 287 degrees in the opposite direction to that of (b). (d) In which direction does earth actually turn?

4. Calculate the following: (a) The number of hours in 360 degrees; (b) The number of earth surface miles in 360 degrees; (c) The distance traveled by Greenwich, England in 24 hours; (d) In 13.5 hours.

5a. Why are the longitudinal lines designated in units of 15 degrees? (b) How many earth surface miles are in 15 longitudinal degrees?

6a, b, c. Reference problem 3: How many miles did earth turn through in parts a, b, c?

7a, b, c. Reference problem 6: How much time passed during each of these rotations of earth? (Assume all answers to be positive.)

A Section of a Cylindrical World Map Showing Entry and Splashdown

This section of a Cylindrical World Map shows the area east of the International Date Line into which your Apollo CM entered and dropped into the Pacific Ocean. Once the time of departure from the moon is established, NASA can then determine your place of entry. This is because the approach path is along the Latitude 10 degrees and due to the time of entry. This section of the Pacific Ocean provides a vast landing area, free of island obstructions over a great distance, and permits a couple of hours of pre-emptive and delay time within the parameters of a fixed schedule. Note that the position of entry is marked as well as the landing position of the CM. Looking carefully at your splash down position, it would appear that you are unnecessarily close to Hawaii. If you measure carefully, you will find that your landing point is approximately 15.4 degrees from Hawaii. This diagonal distance is about 1075 miles and is not close to the islands. However, the distance is much closer to the Johnston Islands. You should check this map section against a complete map of the world for a better understanding of the location.

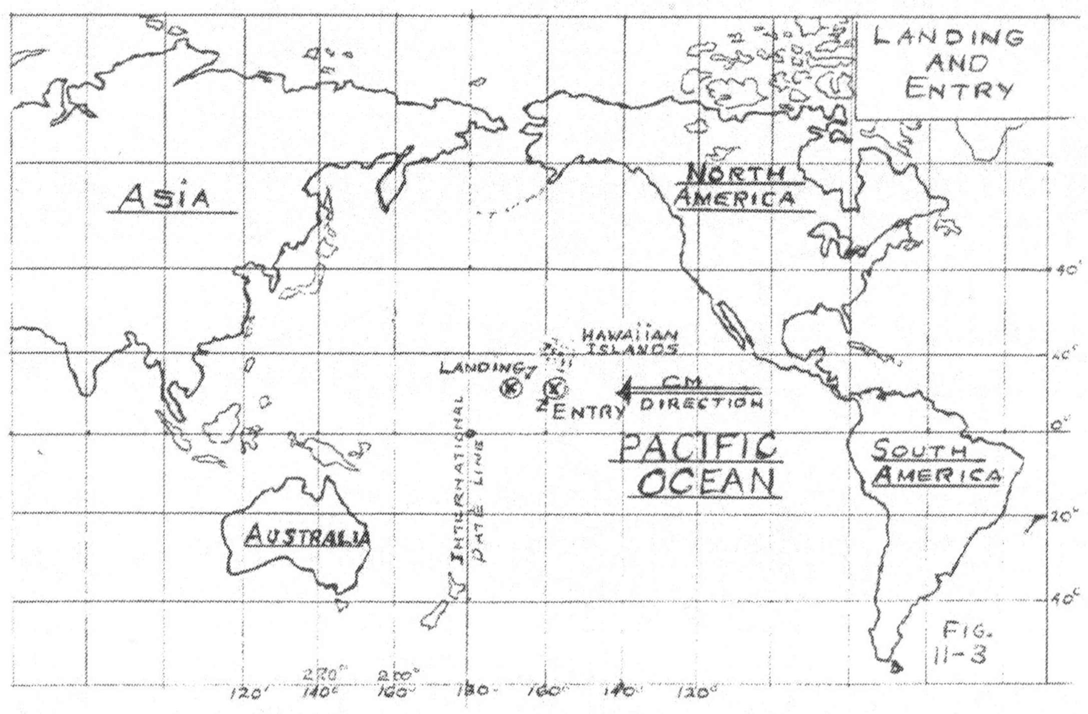

Exercise 47

Self Test

All problems are based upon standard time. Reference the earth chart on page (158). Use the fixed outer chart to determine your answers. Think of the inner circle as a rotating earth. It would be to your advantage to cut out and mark a center disk that you can turn within the outer circle. By doing this, you can subtract or add degrees by rotating the inner chart clockwise or counter-clockwise. All answers must be given in the time frame and velocity mode stated in the problem. Unless told otherwise, you leave earth in its direction of rotation and return in the same direction. Show all required calculations. (Assume that all travel is along the same latitude. Also, assume that angular answers are those taken from the fixed outer circle and that longitudinal answers are those found on the inner, rotating circle.)

(1) If the time at KSC is 6:36 PM, EST, on June 30th, and its earth location is at 81 degrees, what is the time in hours and minutes, and what is the date at KSC, when the earth has rotated (a) 145°? (b) 322°? (c) 450°? (d) 945°?

(2) Your space ship lifts off form KSC at 12:00 noon, on May 10th, and returns at 7:00 PM, on May 15th. The pilot the ship down at what he thinks is the exact location form which it left earth. (a) What was your angular location at the point of lift-off? (b) At what angular location did you land? (c) What angular distance separates your ship from KSC? (d) How many miles is this? (e) How many miles did the KSC rotate during your period of travel?

(3) Your orbital vehicle lifts off from KSC with an average velocity of 18,000 f/s. It lands 8.5 hours later. (a) What linear distance did your ship travel? (b) What is the angular location of your ship

at the landing point? (c) What is your angular distance from KSC? (d) What is your linear distance from KSC?

(4) Reference problem (3): This problem is the same, except that your direction of travel is opposite to the rotational direction of earth.

SPLASHDOWN

As Apollo 11 began its approach to earth, the following information exchange took place between the crew of Apollo 11 and NASA:

"Apollo 11 is 74,906 miles form earth—ship's velocity is 6,854 ft/sec"[17]

"Ship's velocity is 17,322 ft/sec, one hour away from separating the service moduel."[18]

"7,512 nautical miles form earth, ship's velocity is 20,304 ft/sec."

"3,896 miles form earth, ship's velocity is 24,232 ft/sec."[19]

"800 nautical miles form earth, ship's velocity is 33,000 ft/sec."[20]

Apollo 11 landed 241 miles from Johnston Islands and 906 miles form Pearl Harbor.

<u>CONGRATULATIONS – You are Down</u>

Congratulations! You are floating in the Pacific Ocean. A Navy helicopter is flying overhead and Naval divers are already opening the hatch to let in fresh air and to take all of you aboard ship. Tomorrow, your names will be in every newspaper in the world and you will be the envy of all knowing people on earth.

Donald C. Lundy

Where Do You Go from Here

After the recognition and notoriety, after the excitement has passed, after you are no longer known on the street, where do you go? What dreams do you dream? What new frontiers lie ahead for you when you are young and your life is yet to be lived? The Russian space station, Mir: the only permanently orbiting satellite-the only space object permanently manned-has been up there for ten year. The United States and Russia are now jointly building the next generation space station that will orbit 220 miles above earth. Mir is 101 feet long. The length of the new International Space Station will be that of two football fields. It will be manned by six astronauts who will work in seven laboratories. Have you made plans to be part of this project?

Once work is finished on the new International Space Station, plans will be made to launch several tandem space ships to Mars for a first manned landing on that planet. Such a trip will take 220 days travel there and another 220 days back to earth. That makes your trip to the moon look like a "walk around the block" by comparison. This trip must be made from an orbiting platform, as the immensity of such a trip requires too much energy for it to begin on earth. Is there a future for you in this endeavor?

There is presently very little talk about it, but a permanently manned space station on the moon is within the realm of possibility. The technology is already available for this undertaking. The current philosophy is that of "one thing at a time," as it should be. A Moon Station is of great importance to science, because it offers a strategic locale for "future manned flights to other places in our solar system," and for "easy access upon return." As you can see, there is much left to be done. You are at the threshold of a new and fascinating era of discovery. I am sure you will want to be part of these great historic events.

ANSWERS TO EXERCISE PROBLEMS

Exercise (1)

2a. 8.138 days; b. 12:50 PM
5. 186,000 mi/sec
6. 1.01 x 10^{10}mi

Exercise (1A)

	A	B
Aristóteles	72.38 mi	113.65 mi
Eudoxis	51.7 mi	81.15 mi
Serenitatis	465.3 mi	730.55 mi
Manilius	51.7 mi	81.15 mi
Tranquilititis	589.38 mi	925.35 mi
Agrippa	51.7 mi	81.15 mi
Albaategnius	113.74 mi	178.55 mi
Allagensis	77.55 mi	121.75 mi

Exercise (2)

1. 237,087 mi
2. Half
3. 50 minutes each morning
4. 27.33 days
5. Diagram

7a. Flat = 760 mi; b. Sphere=1193.6; ◄
c. The same, 36.33%

Exercise (3)

7. 6.45%
8. Saturn = 8.872 x 10 mi
9. Jupiter – 89,520 mi
10. Mercury = 11.84 f/s-s
11. Mercury = 1406.4 hrs.
12. Mercury dia = 4864 km

Exercise (4)

1a. 2.37 x 10^5 b. 9.57 x 10^{-4}
2a. .00972; b. 634,100
3a. 10^{-4}; b. 10^{-3}; c. 10^{12}
4a. 1.0075 x 10^7 b. 2.059 x 10^{-2}

5. $$\frac{1.7045x10^1}{9.71x10^0} = \frac{9.73x10^2}{5.937x10^3}$$

Exercise (5)

1. 15.37
2. 2.72
3. .388
4. 19.9
5. 9.03
6. 192.55
7. 1.43
8. 3.296
9. s. 616, .755, .781
   c. .788, -.656, .174
   t. .781, -1.15, -5.67
10. 34.19°, 55.8°, 29.34°
11. 6.4 x 10^{-3}
12. 44,697
13. 6.331 x 10^{-4}
14. 8,359,231

Exercise (6)

1. D C

D	C
1	3.14
2	6.28
3	9.42
4	12.56
5	15.7
6	18.84
7	21.98
8	25.12
9	28.26
10	31.4

2. Continued on next page.

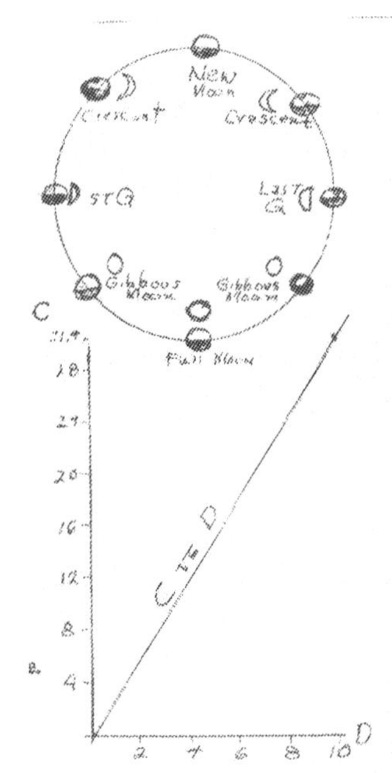

Exercise (6) Continued (2)

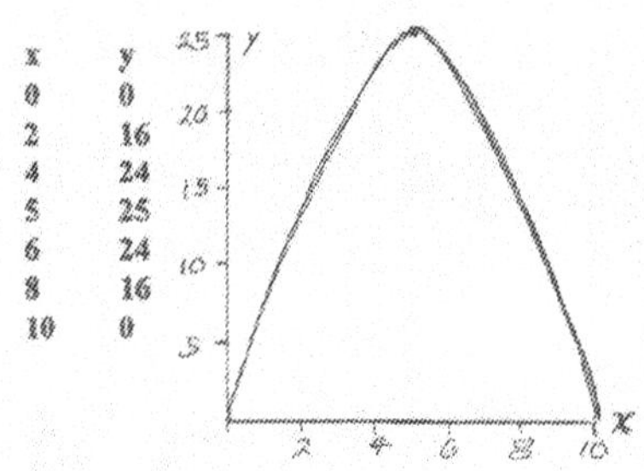

Exercise (7)

1a. 4.33; b. -.999; c.133
2a Fy = 11.57 N; b. Fx = 13.79N
3a. R = 22.02p; b. 72.69°
c. E = 22.02p at 287.31°
4a. Fx = 33.1p;, Fy = 25.9p
b. E = 42p at 0 = 29.58°
5a. 42.54 ft; b. 0 = 29.58°
6. B = 27.13 ft
7. P = 17.32m
8. R = 66.14 at 0 = 146.8°

Exercise (8)

1a. 3000ft; b. .568mi;
c. .909km, d. 909m
2a. 29.33 ft/s; b. 30mi/hr; c. 50 mi/hr
3a. 16,000mi/hr; b. 23,467 ft/s;
c. 25,600km/hr
4a. 24,000mi; b. 38,400km
5. 3,510 mi/hr

Exercise (9)

1a. 225ft/s; b. 1687ft
2a. 3.23ft/s-s; b. 27.3s
3a. 12ft/s; b. 24ft/s; c. 36ft/s
d. 72ft/s; e. 108ft/s
4a. 110ft/s; b. 120mi/hr;
c. 550ft; d. .229mi

Exercise (10)

1a. .707sec; b. 103.7ft; c. Yes
2a. 31.9m; b. 0m/s; c. 2.55sec;
d. 5.1sec
3a. 855.1m/s; b. 86,201m;
c. 2500m/s; 110°

Exercise (11)

1. 980cm/s-s
2. 32f/s-s
3. 62.5 slugs
4. 2144p
5. 9.69kg
6. 3.13m/s-s
7. 3.84f/s-s
8. 60p
9a. 192.8p; b. 229p
c. Yes; force down ramp

Exercise (12)

1a. 7,85ft; b. 3rev; c. 23.55ft/s
d. 1080°; e. 18.84radians
f. 3rev/sec; g. .33sec
h. 23.55ft/sec; i. 443.7ft/s
2a. 2rev/s; 12.57rad/s; b. 1rev/s

Exercise (13)

1a. 30:14:24; b. .033rev/s
c. 147,580km; d. 4,880.3km;hr

Exercise (14)

1. 3,152p; Yes
2. 22.69r/s-s

Exercise (15)

1. 530 Joules
2a. .13; b. .1; c. 8,250f-p
3. 14.67ft
4a. 60s; b. 20s
5a. 383p; b. 321.4p; c. 134.1p
d. 187.3p

Exercise (16)

1. 500,000f-p
2. 467.2f-p
3. Derive
4. 316.3f-p
5. 178.9f/s
6. 300,000Joules

7a. 230,000f-p; b. 13.94HP

Exercise (17)

1a. M1 = 3437.5slug-f/s
M2 = 0slug-f/s
b. 11f/s; c. 3437.5slug-f/s
d. Yes
2. 45 fs
3a. 1.85m/s-s; b. 138,900kg-m/s
c. 138,90N-m; d. 9260N
4a. 7.92 x 10^8; b. 4.302 x 10^9
c. 5.095 x 10^9
5a. 120,000p-s; b. 9813f/s;
c. 11,029f/s

Exercise (18)

1. $\frac{x^2}{a^2}+\frac{y^2}{b^2}=1$

2a. Lx = 2a; b. Ly = 2b
3. F=/a-b
4. $\frac{x^2}{7^2}+\frac{y^2}{5^2}=1$

5a. 29.34f/s-s; b. 9.88f/s
6a. 8.95m/s-s; b. 3.01m/s-s
7a. 354.3mi; b. 34.1km
8[a]. 25,310mi/hr; b. 1.047Hr
c. 28.75f/s-s

Exercise (19)

13. 1.9
14. 285s
15. 9,120f/s

Exercise (20)

StageI: Rm=2.2; Rburn=1.831
StageII: Rm=1.545; Rb=2.83
StageIII: Rm=1.113; Rb=9.89

Exercise (21)

1. Data chart and graph
2. 520,000ft or 98.48mi
3. Results are the same.

4a. Calculation; b. 370,000ft or 70.17mi

5. 28.31mi
6. Sy=53.75mi, Sx=45.1mi

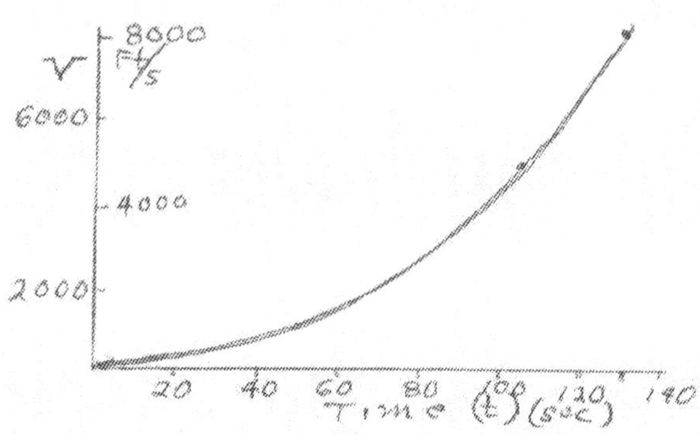

Exercise (22)

1. 6613.33f/s
2. 5528.9f/s
3. 46.55mi
4. 31.26f/s-s
5. 73.16mi

6a. 1258f/s; b. 6787f/s at 55deg
c. 5528.9f/s

7. 8400p

Exercise (23)

1. 25,074f/s	7. 20,533f/s
2. 27.69f/s-s	8. 2000p
3. 8.75s	9. 1161p
4. 11,200f/s	10. 9.11f/s-s
5. R=1.5	
6. 4541f/s	

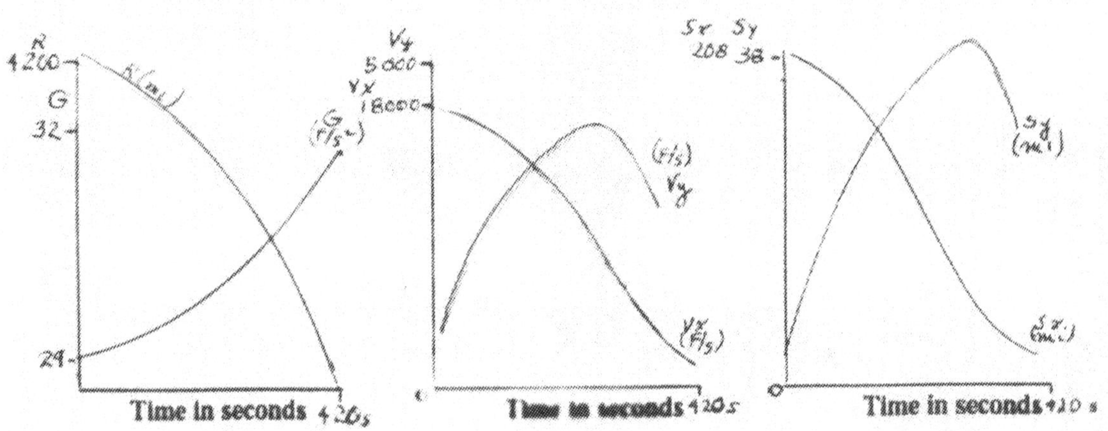

Exercise (25)

G=29.82f/s-s; R=4171mi.
Vy=3327f/s; Vx=14,392f/s
ΔSx=143mi; ΔSy=27.79mi
Riii=4143mi

Exercise (26)

7. (981/37) + (267/43)

Exercise (27)

```
10 LPRINT "Devise a program that multiplies'
20 LPRINT "two variables and adds a third"
30 X=35.1 'This is not in the printout
40 Y=6.33
50 Z=82.7
60 S=X*Y+Z
70 LPRINT "X   Y   Z   S"
80 LPRINT X,Y,Z,S
90 STOP
```

Devise a program that multiplies
Two variables and adds a third

X	Y	Z	S
35.1	6.33	82.7	304.883

Exercise (28)

```
10 LPRINT "Program I"
20 LPRINT "Atlas";,
30 LPRINT "Titan";,
40 LPRINT "Vanguard";,
50 STOP
```

Program I
Atlas   Titan   Vanguard

```
10 LPRINT "Program II"
20 LPRINT "Atlas Titan Vanguard"
30 STOP
```

Program II
Atlas Titan Vanguard

```
10 LPRINT "Program III"
20 LPRINT "Atlas"
30 LPRINT , Titan"
40 LPRINT,,"Vanguard"
50 STOP
```

Program III
Atlas
    Titan
        Vanguard

Exercise (29)

Use the book illustration

Exercise (30)

```
10 LPRINT "Exercise 30"
20 LPRINT "This is a program to find"
30 LPRINT "the perimeter of a polygon"
40 DATA 5, 7, 12.5, 9.0, 3.1
50 READ K, L, M, N
60 P = K + L + M + N
70 LPRINT "K   L   M"
80 LPRINT K, L, M
90 LPRINT "N   P"
100 LPRINT N, P
110 STOP
```

Exercise 30
This is a program to find
The perimeter of a polygon

K	L	M
5.7	12.5	9
N	P	
3.1	30.3	

Exercise (31)

Self Check

Exercise (32)

(1 and 2) Graphing

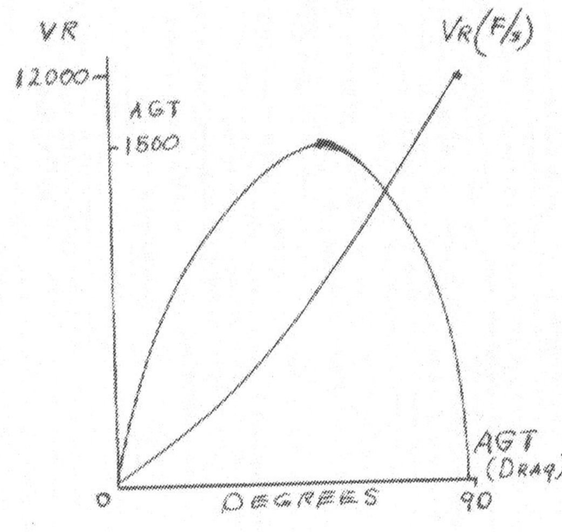

Exercise (32) Continued

3a. Vr vs θ Hyperbolic;
b. AGT vs θ Parabolic

4a. 344; b. 508; -1.48 2nd to 1st
c. No

5a. 2275; b. 2980 –1.31 last to 8th
c. Cosθ→0°This reduces G effect

6a. Drag increases & then decreases
b. At 50°; c. Fuel wt to total wt

7. IWB = (WTS –351,000)/9
8. R=WTS/(WTS-(N)(IWB))
9. AGT=N(IT)(32)(Cos(A/57.3))
10. VR=(8030)(log(R)-AGT)

Exercise (33)

1. 50.3°
2. 2. 5,350 mi
3. 30.19f/s-s
4. 17.88f/s-s
5. 63.62f/s-s

Exercise (33B)

1. 25,074f/s
2. 27.69f/s-s
3. 35,460f/s or 14,177mi/hr
4. 10,387f/s
5. 417.9s
6. 14,401f/s
7. 1.05
8. 12,455f/s
9. 29.8f/s-s
10. 2,477mi
11. 4,962
12. 60.1deg

Exercise (34)

2. 408,441p

Exercise (35)

1. 245,000mi
2. Earth, 215,000mi&Moon,30,000mi
3. 130,659mi
4. 49,630mi
5. 148,890mi
6. 81,667mi

Exercise (35) Continued

7. Average days per month
8. Discussion
9. 1,488,906mi, very nearly
10. 3403mi/hr

Exercise (36)

1. Drawing
2. Perigee

3a. 225,000mi;b. 1,413,000mi
4a. 30,097mi/day;b. 1472mi/hr
5a. 30,097mi/day; b. 1254mi/hr
6a. 42,107mi/day;b. 1554.5mi/hr
7. Apogee (slower velocity)
8. Inverse

Exercise (37)

1. 2.97hr
2. 2.76hr

Exercise (38)

1. 5384f/s
2. 3571mi/hr
3. 55.27mi

4a. 4.96f/s-s;b. 4.61f/s-s
c. .35f/s-s; d. Same
5a. 1500mi;b. .215 orbits
6. Vy=449.7f/s; Vx=5237f/s
7c. R=9259f/s at 56deg
8. Slowing down the Vx component

Exercise (39)

1a. M. Humorum, Lat(15°-30°)S
Long(27°-40°)W
b. M.Tranquil.Lat(3°-37°)N
Long(15°-46°)E
c. M.Nobium Lat(10°-27°)S
Long (10°-20°)W
d. Sinus Iridum(40°-50°)N
Long(25°-37°)W
2a. C. Schickard 45°S. Lat.
2°W Long.
b. C Purbach 26°S Lat
2°W Long.

Exercise (39) Continued

c. C. Atlas 45°N Lat.
47°E Long.
d. C. Copernicus 10° N Lat.
20°W Long.

Exercise (40)

1a. C. Ptolemais 97.6 mi; C. Purbach 73.7mi
b. M. Nectaris 231.2 mi
c. C. Gesendi 69 mi
d. C. Schichard 143.6 mi
2a. Ptol – Purb 284mi
b. Wlm – Tycho 132.4mi
c. Herc – Atlas 68.9 mi

Exercise (41)

1a. 5361f/s or 3655mi/hr
b. 7266mi (P-1.99hrs)
c. 19,618mi
d. 7:52:01AM
e. 342°

Exercise (41B)

1. 13,033p
2. 60s; 10°
3. 738p
4. 7129p

5a. 5741f/s; b. 5653; c. 1000
d. 93mi; e. 58mi
6. Integer
7. Ascending
8. 354f/s
9. Discussion
10. Discussion

Exercise (42) Data Chart

Angle	VRf/s	Vxf/s	Vyf/s	Sx(mi)	Sy(mi)
10°	884	154	871	1.13	6.42
20°	2301	787	2162	5.8	15.92
30°	3881	1941	3361	14.3	24.8
40°	5592	3594	4283	20.5	31.6
50°	7473	5724	4804	42.2	35.4

Gravity/angle=5.33;5.23;5.12;4.98;4.85f/s-s
Allow for considerable variation due to rounding (In R).

Exercise (43)

1. Diagram
2. a=8.8f/s-s;t=16.7s. Time
   for the last 3 miles to the CM.

Exercise (44)

1. 5432f/s or 3704mi/hr
2. 7078mi
3. 1.911hrs
4. 3225f/s=Vav

5a. .201 of orbit; b. 252.4deg
c. 7:10:26AM

6. 150mi
7. 270.7deg

8a. 193.3deg; b. 5556.2mi;
8:40:26AM

9. Chart
10. Drawing

Exercise (45)

1. 2515mi/hr or 3689f/s

2a. 7822f/s or 5333mi/hr;
b. 5333f/s; c. Same

3a. 5361f/s; b. 2461f/s

4. 7183p
5. 130.6s
6. 2422f/s
7. 39f/s
8. 18.55f/s-s
9. 233mi

10a. 3169mi, same; b. 3169mi,same
c. 3402mi, same

Exercise (46)

1. No
2. 252,710mi
3. 2162 mi/hr
4. 121.4mi
5. 163.9s
6. 1154mi
7. 2436f/s
8. 30,000mi
9. 3289.4f/s
10. 11.61hrs
11. Earth pull = moon pull

Exercise (47)

1. Drawing
2. 221,463mi
3. 2515mi/hr
4. 27,118mi
5. 10.49hrs

6a. 166,971mi; b. 2.77days;
c. Same

7. 277,354mi
8. $\theta = 37.01$deg
9. Label the drawing
10. 142.99deg

Exercise (48)

1a. From E→W; b. Counter clock

2. Greenwich, England

3a. 55°/305°; b. 105°/50°; c. 287°/123°

4a. 24hrs; b. 25,120mi; c. Same
d. 13.5hrs

5a. 360deg; b. 1046.67mi

6a. 3837mi; b. 7327mi; c. 20267mi

7a. 3.67hrs; b. 7hrs; c. 19.14hrs

FOOTNOTE REFERENCES

[1]Armstrong, Collins, Aldrin, First on the Moon, 434 pp. (Little Brown and Company, Boston. Toronto, 1970), p. 195.

[4]Ibid., p. 233.
[5]Ibid., p. 197.
[6]Ibid., p. 230.
[10]Ibid., p. 312.
[11]Ibid., p. 314.
[12]Ibid., p. 316.
[13]Ibid., p. 316.
[14]Ibid., p. 319.
[15]Ibid., p. 318.
[16]Ibid., p. 319.
[17]Ibid., p. 348.
[18]Ibid., p. 352.
[19]Ibid., p. 354.
[20]Ibid., p. 355.

[2]Gregory Kennedy, Apollo to the Moon, (Chelsea House Publishers, New York. Philadelphia, 1992), pp. 45, 46.

[3]Ibid., p. 47
[7]Ibid., p. 48.
[8]Ibid., p. 48.
[9]Ibid., p.59.

BIBLIOGRAPHY

Aldrin, Edwin E., Return to Earth, 338 pp., Random House, New York, 1973.

Armstrong, Collins, Aldrin, First on the Moon, 434 pp., Little Brown and Company, Boston . Toronto, 1970.

Asimov, Issac, The Universe – From Flat Earth to quasar, 315 pp., Avon Books, New York, 1968.

Berman, Louis and Evans, J.C., Exploring the Cosmos, Second Edition, Little Brown and Company, Boston . Toronto, 1977.

Branley, Franklyn, Exploration of the Moon, 146 pp., The Natural History Press, Garden City, New York, 1966.

Chartrand, Mark, Exploring Space, 160 pp., Golden Press, new York, 1991.

Kennedy, Gregory, Apollo to the Moon, Chelsea House Publishers, New York and Philadelphia, 1992.

Perry, Greg, Q Basic, By Example, Prentice Hall Computer Publishing, 1993.

Schneider, David, Handbook of Basic, Third Edition, Brady Books, a Division of Simon & Schuster, Inc., 1988.

Sears, Francis and Zemansky, mark, college Physics, Third Edition, 1024 pp., Addison-Wesley Publishing Company, Reading, Mass. And London, England, 1960.

Shephard, Alan and Slayton, Deke, Moon Shot, 383 pp., Turner Publishing, Inc., Atlanta, 1994.

Von Braun, Wernher, The Mars Project, 91 pp., University of Illinois Press, Urbana, 1962.

Weber, R. L., White, M.W., and Manning, K. V., Physics for Science and Engineering, 640 pp., McGraw-Hill Book Company, New York . Toronto . London, 1959.

White, Harvey, Physics An Exact Science, 597 pp., D. Van Nostrand Company, Princeton, N. J. . New York . Toronto . London, 1959.

About the Author

Now retired from teaching, the author taught physics, mathematics, and astronomy in the San Bernardino Secondary School District for thirty years. During that time, he taught physics at San Bernardino Valley College for twenty years. During the last ten years of his teaching, he developed and taught a course in Space Physics in which class members participated in a mathematical trip to the planet Mars. The trip involved calculations required to build and equip an orbiting space station, which was used for the departure point on an intercepted trip to Mars. They did the calculations to determine the proper time to leave the space station in order to intercept the planet and the time of space travel to make orbit and the calculations necessary to make a safe landing on its surface.

After a few years into retirement the author began writing his book ***Lunar Landing and Return***. He chose the Apollo 11 Saga because it was mankind's first successful attempt at making a landing on an extra-terrestrial body and safely returning to earth. Apollo 11 was singularly the greatest achievement in exploration ever achieved by mankind and , all factors considered, might not be achieved successfully today with all of our new technology.

www.ingramcontent.com/pod-product-compliance
Ingram Content Group UK Ltd.
Pitfield, Milton Keynes, MK11 3LW, UK
UKHW061831190726
13855UKWH00005B/1748

9 780759 618589